Graphs and Networks

AN INTRODUCTION

W. L. PRICE, Ph.D.

LONDON BUTTERWORTHS

THE BUTTERWORTH GROUP

ENGLAND

Butterworth & Co (Publishers) Ltd
London: 88 Kingsway, WC2B 6AB

AUSTRALIA

Butterworth & Co (Australia) Ltd
Sydney: 586 Pacific Highway Chatswood, NSW 2067
Melbourne: 343 Little Collins Street, 3000
Brisbane: 240 Queen Street, 4000

CANADA

Butterworth & Co (Canada) Ltd
Toronto: 14 Curity Avenue, 374

NEW ZEALAND

Butterworth & Co (New Zealand) Ltd
Wellington: 26–28 Waring Taylor Street, 1
Auckland: 35 High Street, 1

SOUTH AFRICA

Butterworth & Co (South Africa) (Pty.) Ltd
Durban: 152–154 Gale Street

First published in 1971

© Butterworth & Co (Publishers) Ltd, 1971

ISBN 0 408 70199 4 Standard
 0 408 70200 1 Limp

*Filmset by Keyspools Ltd, Golborne and printed in England by
C. Tinling & Co Ltd, London and Prescot*

GRAPHS AND NETWORKS

OPERATIONAL RESEARCH SERIES

General Editor
K. B. Haley, Ph.D
Professor of Operational Research in Engineering
  Production — The University of Birmingham

This series of books on Operational Research covers in depth practical considerations of the most important techniques. Too often in the past texts have appeared which are largely theoretical in content. The aim of this series is to present the material in a form which can be readily used by practitioners. Each book will be written by an expert in the practical and theoretical aspects of his subject. A mathematical knowledge which would be obtained by a good school leaver is assumed and any more advanced methods that may be required will be developed. Currently in preparation are books in the fields of inventory, simulation, queueing, graphs and networks, replacement and dynamic programming, and it is planned to extend the series into a number of other fields including mathematical programming. The series will be equally suited for individual reading or for use as text books on organised courses.

Preface

This book is an introduction to graphs and networks. In particular it is suitable for final year undergraduate and postgraduate students in Operational Research or related fields. The complete book is suitable for a course of about fifty lecture hours, or Chapters Three and Four (together with some material from Chapter Two) for a thirty-five hour course.

The first chapter sets out why graphs and networks are of interest. To some extent I have anticipated definitions (for example *node* and *arc*) which are given later in the book and have counted on the reader's intuitive understanding to make my points.

The second chapter is meant to serve as a primer in graph theory and a dictionary of definition for the non-mathematician. The reader who is more interested in applications than theory can rapidly cover this chapter and proceed to the rest of the book providing he is willing to return for definitions and theorems as needed.

The third and fourth chapters deal with distance and flow networks respectively. Algorithms for calculating shortest paths and maximum flows are presented, and have been so arranged that block-flow diagrams for a computer code can be drawn up.

Chapter Five is concerned with activity networks and in particular, the project evaluation and review technique (PERT) and the critical path method (CPM). These are the network methods of widest application and their theoretical basis is examined with some of the practical difficulties described.

In addition to the worked examples in the text each chapter is followed by a set of exercises with relevant answers.

I would like to thank those who gave their time to read the manuscript and whose comments brought clarification and improvement. In particular Ray Dignum, Richard Guy, Cecil Law, Jim Moran and Steven Vajda were of great help to me.

I am not sure that a formal dedication is in order in a book of this nature, but in any event I would like to dedicate it to my children Charles and Daniel and Valerie.

Contents

1 INTRODUCTION 1
2 GRAPHS 7
3 SHORTEST PATH PROBLEMS 46
4 FLOW NETWORKS 62
5 ACTIVITY NETWORKS 87
Index 105

CHAPTER 1

Introduction

Graph theory and network theory are branches of mathematics which have grown greatly in the last half century. As early as 1736 Euler published a paper on graphs[4] but books on the subject did not appear until after the first World War[7,11] and wide interest was not awakened until after the second World War (see 1, 2, 5, 6, 10).

This increased activity is accounted for by a need for this kind of knowledge. In recent years we have been building large physical systems which have all the characteristics that mathematicians associate with networks. The road network shown in Fig. 1.1 and the rail network of Fig. 1.2 have little in common in the way of installations and equipment, but a glance at the representations given reveals similarities. Both have centres (or junctions) some joined directly by links. In the links of both networks there is a flow, (cars in one and trains in the other) and in both cases the links can be used to form paths between centres that are not directly connected.

Some transportation networks use more than one type of link. The combined set of railway lines, canals and roads that carry wheat from the Canadian west to the seaports form a grain transportation network. A network of air links, sea links, railways and highways supplying military forces away from their base is described in [9]. In such networks the flow in the links is not vehicles but cargo.

Many problems of sequencing and scheduling can be looked upon as problems in graph and network theory. Perhaps the best known approaches to such problems are the Critical Path and PERT

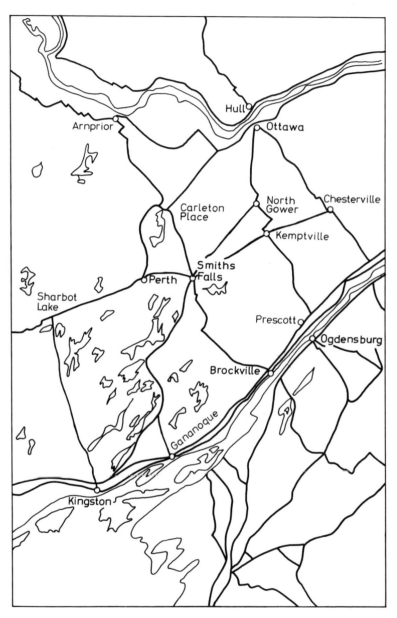

Fig. 1.1

Introduction 3

Fig. 1.2 Railway through routes

methods. Papers describing these techniques started to appear in the late nineteen fifties and since then many books and articles have been published (8 contains an extensive bibliography). CPM and PERT are techniques which permit projects consisting of many individual jobs to be controlled and planned with reference to such aspects as completion date, use of resources, and expenditure of funds.

Some interesting problems in graph theory have been presented in the form of puzzles or games. Probably the first to be published was the Königsburg Bridge Problem[4]. The river Pregel flows through the city of Königsburg[3] (now renamed Kaliningrad) dividing it into four parts as shown in Fig. 1.3. The problem posed

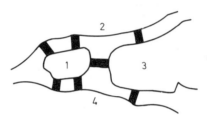

Fig. 1.3. The bridges of Königsburg

was to plan a walk around the city in such a way that each bridge was crossed once and once only. Euler showed that this was not possible. A tour of a network which uses each link once only is called a Eulerian circuit.

A much easier problem would have been to plan a walk such that each of the four sectors of the city was visited once only. Such a tour is called a Hamiltonian circuit after the Irish mathematician W. R. Hamilton.

Finding a 'knight's tour' by starting from any square on a chessboard and moving a knight in such a way that it lands on every square once only, can also be looked upon as a problem in graph theory. Place a dot in the centre of each square of the chessboard and join each dot to all those other dots that can be reached in one move of the knight. The result is a graph representing all possible moves a knight can make on the board. A Hamiltonian circuit of this graph is a knight's tour.

From these examples we have seen that various kinds of transportation systems, scheduling problems and puzzles have certain common characteristics. When a situation such as this arises it is often profitable to study the common characteristics as a separate field.

The networks that we will study are thus abstractions of real systems and out of all the common elements of such systems we will concern ourselves only with the centres (usually called *nodes*), the links (usually called *arcs*), and one or more *arc functions*, which are arrays of numbers giving some information about the arcs (such as length, time to traverse and so on). Such problems as finding the shortest path between two points in a network, or routing so that the flow between two centres is a maximum may be solved by the use of network theory.

Network theory is used as a framework for the study of those properties of systems which depend both on form and arc function. For example the largest possible flow between node S and node T

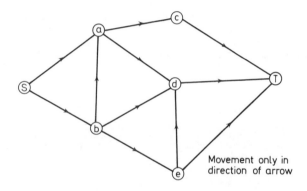

Fig. 1.4. *Movement only in the arrow direction*

in Fig. 1.4 depends on the number of paths between the nodes and also on the capacity of the arcs involved to carry the flow.

Many of the properties of real systems which are of interest to us depend only on form. For example if Fig. 1.4 represented a network of one-way streets, then while it is possible to travel from S to T, the return journey is not possible, nor is there any path from C to E.

Graph theory concerns itself only with the form of a system, that is, the number of nodes and the number and arrangement of the arcs. Just as networks are abstractions of real systems, graphs can be looked upon as abstractions of networks. Leaving aside arc functions, in the chapter that follows we will concentrate on form and attempt to extract all the information about the system which depends on form alone. The remaining chapters deal with networks and specific applications.

6 Graphs and Networks

REFERENCES

1. BERGE, C. et GHOUILA-HOURI, A., *Programmes, Jeux, et Réseaux de Transport*, Dunod, Paris (1962).
2. BUSACKER, R. G. and SAATY, T. L., *Finite Graphs and Networks*, McGraw-Hill, New York (1965).
3. ERHARDT, T., *Geschichte der Festung Königsberg* (PR1257—1945), Wurzburg, Hozzner (1960).
4. EULER, L., 'The Königsburg Bridge Problem', *Publications of the St Petersburg Academy of Science* (1736).
5. HARARY, F., NORMAN, R. and CARTWRIGHT, D., *Structural Models: Introduction to the Theory of Directed Graphs*, Wiley, New York (1965).
6. International Symposium, Rome 1966, *Theory of Graphs*, Dunod and Gordon and Breach (1967).
7. KÖNIG, D., *Theorie der Endlichen und Unendlichen Graphen*, Leipzig (1936).
8. LEVIN, R. I. and KIRKPATRICK, C. A., *Planning and Control with PERT/CPM*, McGraw-Hill (1966).
9. LEWIS, R. W., ROCHOLDT, E. W., WILKINSON, W. L., 'A Multi-Mode Transportation Network Model', *Naval Research Logistics Quarterly*, **12**, No. 327 (1965).
10. ORE, O., *Graphs and Their Uses*, Random House, Toronto (1963).
11. SAINTE-LAGUË, A., 'Les Réseaux (ou Graphes)', *Memorial des Sciences Mathématiques*, **18**, Paris (1926).

CHAPTER 2

Graphs

2.1. Basic Concepts

A *graph* G is a structure having a non-empty set N of elements called *nodes* and a set A of elements called *arcs*. Each element of A is composed of an unordered pair of distinct elements selected from N (a given pair may be selected more than once). A graph is often referred to by the notation $G(N, A)$.

This structure can be represented in the following way. Let each node be represented by a point and let the elements of A be represented by lines joining two points.

The following set of arcs and nodes satisfies the definition of a graph given above (Fig. 2.1):

$N = (a, b, c, d, e, f)$
$A = (ab, ac, bc, be, bf, ef, ed, ed, cd, df)$

If the set A is empty and the graph consists only of nodes, the result is called a *null graph* (Fig. 2.2). If the set A consists of all possible the pairs (each taken once only) which can be chosen from N, then G is called a *complete graph* (Fig. 2.3). If it is known that a graph is complete then all the properties of that graph can be specified merely by indicating the number of nodes. Such a graph is denoted by K with a subscript showing the number of nodes. The graph of Fig. 2.3 is K_6.

Sometimes we are interested only in subsets of the nodes or arcs. For this reason we make the following definitions. A *subgraph* of

8   *Graphs and Networks*

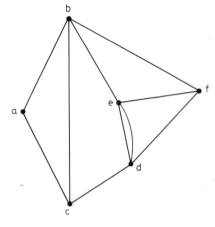

Fig. 2.1

Fig. 2.2. Null graph on six nodes

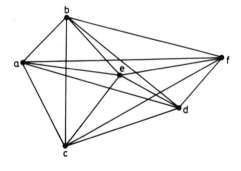

Fig. 2.3. Complete graph (K_6)

G (N, A) is a graph consisting of a subset of the nodes of N and all the arcs of A joining the nodes of this subset. A *partial graph* of G (N, A) has all the nodes of N and some subset of the arcs of A.

In Fig. 2.1 nodes (a, b, c) and arcs (ab, bc, ca) form a subgraph, while the arcs (ab, bf, fd, de, dc) together with all the nodes form a partial graph. The extreme case of a subgraph is one node (no arcs). The extreme case of a partial graph is the null graph on some set of nodes.

The *end points* of an arc are the two nodes which it joins. If an arc does not have distinct end points that arc is said to form a *loop*, or *self loop*. If several arcs have the same end points they are said to be

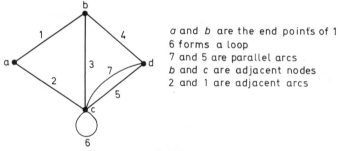

a and b are the end points of 1
6 forms a loop
7 and 5 are parallel arcs
b and c are adjacent nodes
2 and 1 are adjacent arcs

Fig. 2.4

parallel. Two nodes are said to be *adjacent* if they are the two end points of an arc. Similarly two arcs are said to be adjacent if they have one common end point. All these definitions are illustrated in Fig. 2.4.

If it is possible to divide the nodes of a graph into several sets so that no nodes within a set are adjacent the graph is said to be

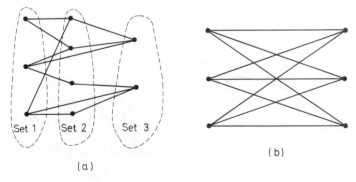

Fig. 2.5

10 Graphs and Networks

multi-partite (Fig. 2.5a). The graph shown in Fig. 2.5b is the complete bi-partite graph on two sets of three nodes where each node of one set is joined to all nodes of the other set.

Note that different pictorial representations arrived at by changing the position of one or more nodes of a graph, and then stretching the arcs so that all existing connections still connect does not alter the intrinsic structure of the graph. If a continuous transformation of this type is made to a graph $G\,(N, A)$, changing it to $G'\,(N', A')$ then $G\,(N, A)$ is said to be *homeomorphic* to $G'\,(N'\,A')$. This is illustrated in Fig. 2.6. Note also that a graph $G\,(N, A)$ is said to be *isomorphic*

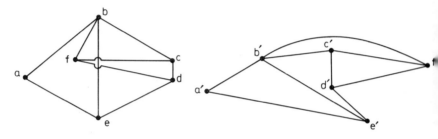

Fig. 2.6

to a graph $G'\,(N', A')$ if each node of N can be identified with a node of N' and each arc of A with an arc of A' in such a way that the incidence matrices of G and G' are identical (see Section 2.6).

In a large complex graph it may be confusing to count the number of arcs directly. However an accurate count can be obtained using the concept of *degree*. The degree of a node is the number of arcs incidence at that node. For example in Fig. 2.6, the degree of a is 2, and the degree of e is 3. In notation we write, $\rho(a) = 2$ and $\rho(e) = 3$. Now if the degree of each node is found then each arc will have been considered twice because of its two end points. Therefore the sum of the degree of all nodes is twice the number of arcs. In Fig. 2.1 the number of arcs is:

$$\tfrac{1}{2}(\rho(a)+\rho(b)+\rho(c)+\rho(d)+\rho(e)+\rho(f))$$
$$=\tfrac{1}{2}(2+4+3+3+4+4) = 10$$

Since the sum of the degree of all nodes is twice the number of arcs, it can be seen that this number will always be even. If we call nodes having an even degree *even* and those having an odd degree *odd*, it is evident that in any graph there is always an even number of odd nodes.

The concept of degree can be carried over from nodes to subgraphs. The degree of a subgraph is the total number of arcs joining some node of the subgraph to some other node outside the subgraph.

We will now consider a class of graph having a property which is useful in modelling certain systems. A *directed graph*[3] is a structure having a set N of elements called nodes, a set A of elements called arcs and each element of A is composed of an *ordered pair* of elements selected from N. This means that the pair ab is not the same as the pair ba.

In pictorial representation the ordering is shown by placing an arrowhead on each arc with the point of the arrow indicating the second node of the pair. For example in Fig. 2.7, the arcs of the

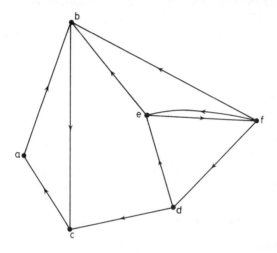

Fig. 2.7

graph are $(ab, ca, bc, eb, de, fb, ef, fe, fd, dc)$. This could represent a one-way street complex, or a communication system where each node represents a centre which sends information to some points and receives it from others. If one centre can both send and receive information it can be indicated by using two parallel arcs in opposite directions (nodes e and f of Fig. 2.7). It can also be indicated by using one arc carrying no arrowhead to join the nodes. In this latter case the graph is said to be partially directed.

Many of the concepts defined in this section can be used for directed graphs without alteration, but there are additions. For example in a directed graph we can define an *indegree* and an *outdegree* at each node and for each subgraph. The indegree of a node is simply the number of arcs having the node as the second endpoint in the ordered pair. The outdegree is the number of arcs

2.2. Chains and Paths

A *chain* (or *simple chain*) is a set of at least two arcs such that:
(1) There are exactly two arcs of the set which are each adjacent exactly one other arc of the set (not necessarily the same arc).
(2) Any remaining arcs of the set are each adjacent to exactly two other arcs of the set.
(strictly speaking a single arc is also a chain joining its own two endpoints).

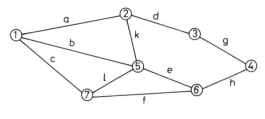

Fig. 2.8

If we use the symbol $\sim$ to mean 'is adjacent to' then in Fig. 2.8,
(1) adg is a chain where $a \sim d, d \sim a, d \sim g, g \sim d$.
(2) cf is a chain where $c \sim f, f \sim c$.
(3) $adgh$ is a chain where $a \sim d, d \sim a, d \sim g, g \sim d, g \sim h, h \sim g$.

If Fig. 2.8 were the graph of a road network where each arc represented a section of highway and each node a city, then the chains of this graph would describe possible routes for travelling from one city to another. This is because the definition obliges the set of arcs to be connected (so that there are no breaks in the route) and prevents three or more arcs of the set from having a common end point (so that there are no forks in the route). The first condition of the definition also prevents the set of arcs from forming a closed figure so that each chain is a route between two distinct nodes.

Note that substituting the word node for the word arc in the definition of a chain still results in a chain. For example in Fig. 2.8, the set of nodes $(1, 2, 3, 4, 6)$ satisfy such a definition and refer to the same route from node *1* to node *6* as the chain $adgh$.

A *cycle* is a set of arcs such that each arc is adjacent to two other arcs of the set. Roughly speaking it is a closed chain. In Fig. 2.8

adgheb, befc, adghfc are cycles. Note that the definition excludes eight-shaped figures such as *aklfeb*. Such figures are called *compound cycles* and defined more precisely as a connected (see the following definition) set of arcs such that each arc of the set is adjacent to an even number of other arcs in the set.

A graph *G* is said to be *connected* if from each node of *G* there is

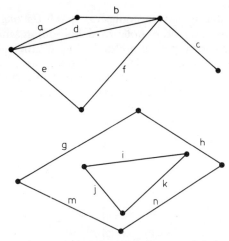

Connected component 1: arcs abdef
Connected component 2: arcs ghmn
Connected component 3: arcs ijk

Fig. 2.9

at least one chain to every other node. The graph shown in Fig. 2.9 is not connected but has three *connected components*.

In Fig. 2.9 if arc *c* is deleted one node becomes isolated. If it is possible to destroy the connectivity of a component by the removal of just one arc that component is said to be not *strongly connected*. A strongly connected graph is one where every arc belongs to at least one cycle. This ensures that there are at least two chains from each node to every other node and the graph will therefore remain connected after the removal of any one arc. The other two components of the graph in Fig. 2.9 are strongly connected.

For a directed graph the definition of a chain remains unchanged but a *path* from any node *i* to any node *j* is defined as a chain from *i* to *j* in which all the arcs are aligned in the same direction, and that *i* is the first endpoint of the first arc of the chain and *j* is the second endpoint of the last arc of the chain. In Fig. 2.7 (*fd, dc, ca, ab*) is a

14   *Graphs and Networks*

path from f to b, but (ab, bf) is not a path at all. Both are chains. In an undirected graph the terms path and chain may be interchanged.

The definition of a *circuit* is derived from the definition of a cycle in a similar way. A circuit is a cycle in which no node is the second (or first) endpoint of more than one arc. In Fig. 2.7 the cycles (ab, bc, ca) and (de, ef, fd) are circuits while (eb, fb, fe) is a cycle but not a circuit.

It is possible to have a connected graph (having a chain from each node to every other node by definition) in which there is no path between some pair of nodes. In Fig. 2.7 the subgraph (a, b, c) has an outdegree of zero. There is therefore no path from a, b or c to d, e or f (note that paths in the opposite direction exist).

The concepts of cycle and circuit are related to the *co-cycle* and the *co-circuit* (this is known as a *dual* relationship and will be explored later). Consider a connected graph G. A co-cycle in G is a

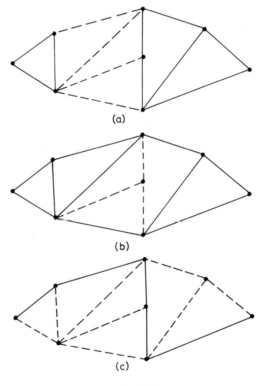

Fig. 2.10

set of arcs which if removed from G would disconnect the graph into more than one connected component. Fig. 2.10 shows several examples of co-cycles and illustrates the fact that the connected components produced are subgraphs having one or more nodes. The co-cycles of Fig. 2.10a and 2.10b are *simple co-cycles* which means that they do not properly contain a smaller co-cycle. The co-cycle of Fig. 2.10c is a *compound-co-cycle* composed of two simple co-cycles each of which isolates one node. The question of adding cycles (or co-cycles) will be dealt with in a later section.

A co-cycle has been defined in terms of arcs which if removed would disconnect the graph. It can also be defined in terms of nodes which we may wish to separate from the graph. Given a graph $G(A, N)$ and a subset N' of the nodes, draw simple closed curves around the nodes on N' so that:

(1) Adjacent nodes of N' are enclosed by the same closed curve;
(2) No node of $\overline{N}'$ (the complement of N') is enclosed by a closed curve.

The set of arcs A' which is intersected by the curves is said to be the co-cycle of N' and by definition it is also the co-cycle of $\overline{N}'$. This is illustrated in Fig. 2.11 ($\omega(N')$ is used as the symbol for the co-cycle of a set of nodes N' and is written:)

$$\omega(N') = \omega(\overline{N}')$$

The definition of a co-circuit is derived from the co-cycle just as the definition of a circuit is derived from the cycle. A co-circuit is a

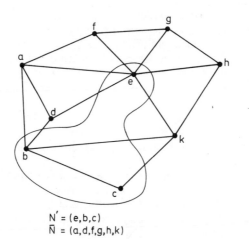

$N' = (e,b,c)$
$\overline{N} = (a,d,f,g,h,k)$

Fig. 2.11

co-cycle in which all the arcs contribute either to the indegree or to the outdegree of some subset of the nodes of the graph.

A *tree* of a connected graph $G(A, N)$ is a connected partial subgraph of G (it contains a subset of the nodes and some subset of the arcs joining those nodes) which contains no cycles. A *spanning tree*

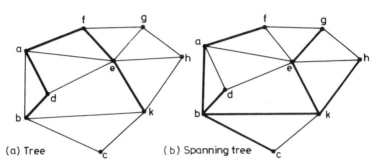

Fig. 2.12

is a connected partial graph (all the nodes and a subset of the arcs) containing no cycles. A tree and spanning tree of the graph in Fig. 2.11 are shown in Fig. 2.12.

There are at least six equivalent definitions of the spanning tree H of a connected graph G having n nodes. Each definition makes use of a different combination of the properties of spanning trees. The definitions are[1]:

(1) H is connected and contains no cycles.
(2) H has *n–1* arcs and contains no cycles.
(3) H has *n–1* arcs and is connected.
(4) H contains no cycles but the addition of one more arc of G creates one and only one cycle.
(5) H is connected and the removal of any one arc causes disconnection.
(6) Each pair of nodes is connected by one and only one chain.

Where G has more than one connected component, one spanning tree can be selected from each connected component and the collection of trees is called a *forest*.

W, a partial graph of G is called a *co-tree* if it contains no co-cycles of G and is such that the transfer of one more arc from $\overline{W}$ ($\overline{W} = G - W$ and is the complement of W) to W would form one simple co-cycle and one only. From this definition it can be seen that there is no requirement for W to be connected.

It will now be shown that if W is a co-tree then $\overline{W}$ is a spanning tree. It is evident that $\overline{W}$ is connected and reaches all nodes for if this were not so, $\overline{W}$ would contain a separating set of arcs (a co-cycle) which is not possible by the definition of W. It can also be seen that $\overline{W}$ contains no cycles, for if it did some arc of a cycle in $\overline{W}$ could be transferred to W without forming a co-cycle in W, which contradicts the definition of a co-tree. Therefore $\overline{W}$ satisfies both criteria of definition (1) of a spanning tree.

The converse is also true. Let H be a spanning tree and $\overline{H}$ its complement. All nodes of G are reached by at least one chain of H and so H contains no isolated nodes. Therefore $\overline{H}$ contains no co-cycles (or H would have more than one connected component). Since there is only one chain between each pair of nodes in H, the transfer of an arc from H to $\overline{H}$ would isolate some node or group of nodes forming a co-cycle in $\overline{H}$. Therefore $\overline{H}$, the complement of a spanning tree is a co-tree.

In Fig. 2.12b a spanning tree is shown by the heavy lines and its complement, a co-tree, is indicated by the lighter ones.

2.3. Cut-sets of Arcs

A *two-terminal graph* (also called a *one-port graph*) is a graph in which two selected nodes are treated differently from the others. One of these is usually called the *source* (S), and the other the *sink* (T). Two-terminal graphs are of interest to us because they mirror the elements of such real networks as transport and communications.

A *cut-set* (or *cut*) is a set of arcs which if deleted will separate a connected two-terminal graph into two connected components, one containing S and the other containing T. In addition, no subset of the arcs of a cut affects such a separation. In effect it is the co-cycle of some set of nodes which contains either the source or the sink.

Cut set	arcs	cut set a co-cycle of:
1	a b	S
2	c d e f	S,1,2
3	c d e i m	S,1,2,5,6
4	l m	S,1,2,3,4,5,6,7

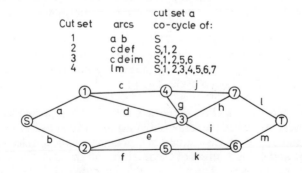

Fig. 2.13

Fig. 2.13 shows four cuts of a graph and lists the nodes of which cut is a co-cycle.

Another definition of a cut can also be made. If S and T are not adjacent nodes, join them with an imaginary arc. A cut is then any co-cycle which includes the arc joining S and T.

Theorem: *In a two-terminal graph $G(A, N)$, every chain linking S to T must cross every cut an odd number of times.*

Let C be any cut. It divides N into two subsets N_1 and N_2 such

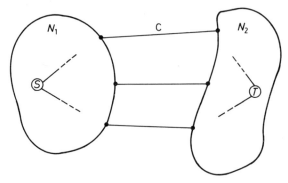

Fig. 2.14

that $S \in N_1$ and $T \in N_2$ (Fig. 2.14). Let c be the number of times that a chain from S to T crosses C. (See page 34.)

If $c = 2p$ (where p is any positive integer) the chain will make p $N_1 \to N_2$ crossings and p $N_2 \to N_1$ crossings. Because the first is from $N_1 \to N_2$ the last will be from $N_2 \to N_1$ and since the chain cannot terminate at T, c cannot be an even number.

If $c = 2p+1$ (this will be a positive odd integer) there will be p $N_1 \to N_2$ crossings and p $N_2 \to N_1$ crossings followed by a last crossing from $N_1 \to N_2$. The chain can therefore reach T and any positive odd number of crossings is permitted.

It can be proved that an S–T chain must intersect every cut an odd number of times and by extending the reasoning, it can be shown that every cycle has an even number (zero being even) of arcs in common with every co-cycle. The proof of this will be left to the reader.

In a directed graph $G(N, A)$ the arcs of a cut C can be divided into two subsets. As before the set N is divided into two subsets such that $S \in N_1$ and $T \in N_2$. C^+ is a subset of C in which all the arcs lead from N_1 to N_2 and C^- is a subset of C containing all the arcs of C leading from N_2 to N_1. Note that it is possible for either C^+ or C^- to be an empty set. Deletion of all the arcs of C disconnects S from T

completely. Deletion of the arcs of C^+ breaks all the chains from S to T and deletion of all the arcs of C^- breaks all the chains from T to S.

2.4. Planar and Non-planar Graphs

A graph which can be drawn on a surface so that no arcs intersect (except at nodes) is said to be *imbedded* in that surface. Fig. 2.15 shows graphs imbedded in a sphere and in a plane.

We are interested particularly in those graphs which can be imbedded in a plane and are called *planar graphs*. Two *non-planar*

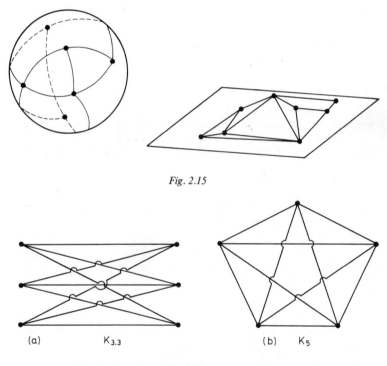

Fig. 2.15

(a) $K_{3,3}$

(b) K_5

Fig. 2.16

graphs are shown in Fig. 2.16. The first is the complete bi-partite graph on two sets of three nodes. This graph (known as $K_{3,3}$) is often called the *utility graph* as it is the problem of connecting three houses to gas, water and electricity without having any of the

20 *Graphs and Networks*

service lines cross. The second (called K_5) is a complete graph on five points.

There is a relation between the number of nodes, the number of arcs and the number of *regions* marked out in a plane by a connected planar graph. Each region (Fig. 2.17) is bounded by k nodes and

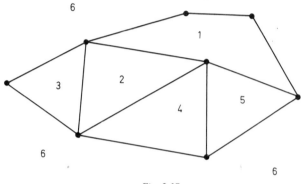

Fig. 2.17

k arcs. Included in the definition of a region is the area of infinite size which lies entirely outside the graph. For all planar graphs having a arcs n nodes and r regions (also called *faces*) we will show that $n-a+r = 2$. This relation obviously holds if $r = 2$ (Fig. 2.18a)

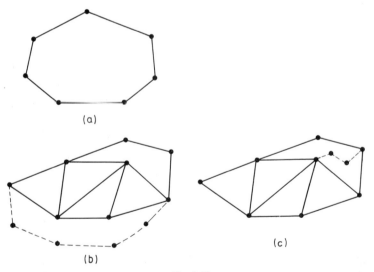

Fig. 2.18

and we will show inductively that if the relation is true for some particular value of r it is also true for any integer value of r.

Assume the existence of a graph having r regions and satisfying the relation $n-a+r = 2$. Add a new region to the graph. If the new region is created by building it on to the graph using some of the infinite region it is evident (Fig. 2.18b) that this will require q new nodes and $q+1$ new arcs. If the new region is created by dividing some existing finite region into two parts (Fig. 2.18c), this also requires q new nodes and $q+1$ new arcs.

Now as the relation holds for the initial graph having r regions substitute in the parameters of the new graph to obtain:

$$(n+q)-(a+q+1)+(r+1) = 2$$

Thus the general expression:

$$\text{nodes} - \text{arcs} + \text{regions} = 2$$

is true for all planar graphs and is called *Euler's Formula*.

It is interesting to note why Euler's Formula does not hold for non-planar graphs. An examination of Fig. 2.16 will show that the regions cannot be determined as we have done above and so the formula cannot be applied.

A second theorem due to Kuratowski shows that the necessary condition for a graph to be planar is that it does not contain a subgraph homeomorphic to K_5 or $K_{3,3}$ as a subgraph.[1] The theorem implies that all non-planar graphs can be broken up into a number of planar subgraphs (the number may be zero) and a number of non-planar subgraphs with each homeomorphic either to $K_{3,3}$ or K_5.

For every planar graph G we can define a *dual graph* G^* in the following way (any set of rules which when applied to a *primal* structure produce a second of the same kind must be reversible if the second structure is to be called the *dual* of the first. Note that the terms *dual* and *primal* are arbitrary and may be interchanged).

(1) In each region of the original graph (most often referred to as the *primal graph*) place a dual node.

(2) Join dual nodes which are in adjacent regions once for each arc forming the boundary separating the regions. (Fig. 2.19)

Theorem: To every cycle of G there corresponds a co-cycle of G^*.

By construction each arc of G corresponds to an arc of G^* and each face of G to a node of G^*. Consider some cycle C of G. Cycle C encloses a set of faces of G each of which is adjacent to at least one other face of the set. Therefore by construction, the set of dual nodes M^* corresponding to these faces are the nodes of a connected subgraph of G^*. Let the set of dual arcs corresponding to the primal

22   *Graphs and Networks*

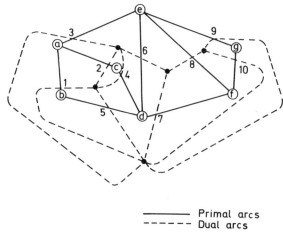

Fig. 2.19

arcs of C be called C^*. It will be shown that C^* is the co-cycle of M^*.

All arcs of the co-cycle of M^* belong to C^*. Suppose that some arc of the co-cycle of M^* does not belong to C^*, that is to say some node of M^* is connected by an arc of $\bar{C}^*$ to a node of $\bar{M}^*$. This latter node corresponds to a face of G which is outside C. To this dual arc must correspond a primal arc which is not of C and at the same time is on the boundary between the set of faces corresponding to M^* and the set of faces corresponding to $\bar{M}^*$ Since this is a contradiction such an arc cannot exist.

All arcs of C^* belong to the co-cycle of M^*. The supposition that there is an arc of C^* which is not of the co-cycle of M^* implies that there is an arc of C whose corresponding dual arc did not have one endpoint inside C and one outside. This too is a contradiction.

Since C is any cycle of G the theorem has been proved.

It can also be shown that the dual of a tree is a co-tree. Let G be a graph with n nodes, a arcs and r regions. The dual of G (G^*) will have a arcs, r nodes and n regions. Any tree H of G has $n-1$ arcs and contains no cycles. The corresponding figure in G^* will therefore contain $n-1$ arcs and no cycles by the preceding proof. Since a tree of G^* has $r-1$ arcs a co-tree of G^* will have $a-r+1$ arcs. By Euler's Formula we have $n-1 = a-r+1$. Therefore the figure in G^* is a co-tree by definition. Since H was any tree of G the statement holds generally.

Defining the dual of a two-terminal graph presents us with certain problems not encountered before. Those chains which identify paths from the source node to the sink node are of particular interest. We would like their corresponding figures in the dual graph to be of interest as well.

The following procedure (which produces the dual figures most often shown in texts) brings about the desired result. Join the source S and the sink T with an artificial arc. (If it is not possible to do this without causing the graph to become non-planar, the graph is not *S–T planar* and the dual with respect to that particular choice of source and sink is undefined) creating another region in the plane. Place dual nodes in each region and derive the dual graph in the same way as previously described. Fig. 2.20 illustrates this procedure.

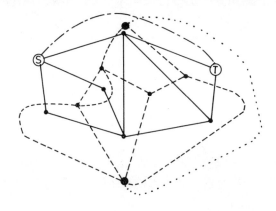

Fig. 2.20

One node is located between the primal artificial arc and other arcs of the primal graph and the other is in the infinite region. Either of these two nodes may be called the dual source whereupon the other becomes the dual sink.

In the dual graph defined with respect to a source and sink, the figure corresponding to a path between S and T in the primal network is a cut separating the dual source and dual sink. The proof of this statement follows directly from the duality between circuit and co-circuit. If one considers the path in the primal network as part of a circuit containing the primal artificial arc then by duality, the corresponding figure is a co-circuit containing the dual artificial arc which if removed, leaves us with a cut in the dual network.

If the primal network being considered is directed the dual network must also be directed if a dual relationship is to be preserved. A convention must be adopted for orientating the arcs of the dual network. This need only be a simple left or right hand convention where the arrow indicating the direction of a primal arc is rotated

24 *Graphs and Networks*

clockwise (or anti-clockwise) until it coincides with the corresponding dual arc. This is illustrated in Fig. 2.21.

However this is not sufficient to ensure the duality of all concepts. Note that in Fig. 2.21, arcs *a* and *e* form a cut in the directed primal

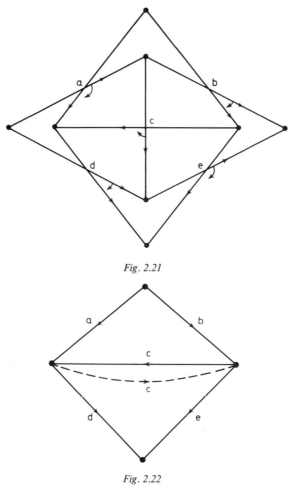

Fig. 2.21

Fig. 2.22

network but they do not correspond to a cut in the directed dual network. This difficulty can be overcome in the following way. Replace all directed arcs in the dual (with the exception of those leaving the source or entering the sink) by two parallel arcs. One of these arcs has the direction given by the convention just described and an arc length described in the usual way. The other is given the opposite direction and has an arc length of zero (Fig. 2.22). In this way all cuts in the primal network can be made to correspond to paths in the dual.

2.5. Matrix Representations of a Graph

Up to this point we have used pictorial representation for a graph. This representation is useful for showing the connection between some real system and its graphic model but may not be convenient to manipulate. Computers can store and manipulate numbers easily but they cannot directly store pictorial information.

Several different matrix representations have been devised, and three of them will be described in this section. Which one is used depends on the application.

2.6. The Incidence Matrix

The incidence matrix E of a graph G is formed in the following way. A column of the matrix is allocated to each arc and a row to each node. An entry e_{ij} is *1* if node i is an endpoint of arc j and is zero otherwise. The incidence matrix for the primal graph of Fig. 2.19 is shown below.

	1	2	3	4	5	6	7	8	9	10
a	1	1	1	0	0	0	0	0	0	0
b	1	0	0	0	1	0	0	0	0	0
c	0	1	0	1	0	0	0	0	0	0
d	0	0	0	1	1	1	1	0	0	0
e	0	0	1	0	0	1	0	1	1	0
f	0	0	0	0	0	0	1	1	0	1
g	0	0	0	0	0	0	0	0	1	1

Obviously where G has A arcs and N nodes the dimension of this matrix is $A \times N$.

It is necessary to know that the incidence matrix of some graph G can be stored away and reproduced at some future date. Let the graph constructed from E be G'. We will now see that G' must be isomorphic to G and that therefore E can be used without error to provide a true record of the graph.

The only nodes which can exist in G' are those of G because no others exist in E. We are also assured that G' will have the correct number of arcs because the number of columns in E is exactly equal to the number of arcs in G. To show now that G' is isomorphic to G, it is sufficient to demonstrate the impossibility of some node i of G' having an arc j incident to it when the corresponding situation does not occur in G. Now if arc j were incident to node i in G' it would imply that $e_{ij} = 1$. If this is so it is because arc j is incident to node i in G. Therefore G' is isomorphic to G which means that the translation from E to G is unique.

Before continuing the discussion of the incidence matrix it is useful to define *modulo 2 addition*, and the *ring sum* of two rows of a matrix.[4] The sum of two integers modulo 2, is equal to the remainder which is obtained when their arithmetic sum is divided by two. For example:

$$1+0 = 1 \bmod 2$$
$$1+1 = 0 \bmod 2$$
$$1+2 = 1 \bmod 2$$
$$1+3 = 0 \bmod 2$$
$$3+4 = 1 \bmod 2$$

$R_1 \oplus R_2$, the ring sum of rows R_1 and R_2 of a matrix is the new row obtained when the corresponding elements of R_1 and R_2 are added modulo 2. For example:

$$R_1 = (1,0,1,0,0,1,1)$$
$$R_2 = (1,0,0,1,1,1,0)$$
$$\text{then } R_1 \oplus R_2 = (0,0,1,1,1,0,1)$$

It can be shown that the ring sum satisfies the associative rule $R_1 \oplus (R_2 \oplus R_3) = (R_1 \oplus R_2) \oplus R_3$ and the commutative rule $R_1 \oplus R_2 = R_2 \oplus R_1$.

Looking again at the incidence matrix of the graph in Fig. 2.19, it is easy to verify that the ring sum of any six rows is equal to the seventh row. This is not chance or coincidence but a general characteristic of incidence matrices. If a connected graph has n nodes then the incidence matrix has $n-1$ *independent rows* (excluding *self-loops*). A set of independent rows is said to form the *basis* of the matrix because from them the whole matrix can be generated.

It is not difficult to see why only $n-1$ rows of E should be independent. By the time that $n-1$ nodes have been considered both end points of some arcs have been located and therefore, the entries for these arcs in the final row must be zero. Similarly only one end point of other arcs will have been located and the entry for these in the final row must be one.

Directed graphs also have an incidence matrix representation. For a directed graph, $e_{ij} = +1$ if arc j is incident to node i and contributes to its indegree and -1 if arc j is incident to node i and contributes to its outdegree. Again, $e_{ij} = 0$ if arc j is not incident to node *i*. The incidence matrix of the directed graph shown in Fig. 2.23 is given below.

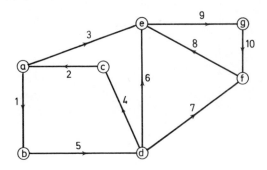

Fig. 2.23

	1	2	3	4	5	6	7	8	9	10
a	−1	1	−1	0	0	0	0	0	0	0
b	1	0	0	0	−1	0	0	0	0	0
c	0	−1	0	1	0	0	0	0	0	0
d	0	0	0	−1	1	−1	−1	0	0	0
e	0	0	1	0	0	1	0	1	−1	0
f	0	0	0	0	0	0	1	−1	0	1
g	0	0	0	0	0	0	0	0	1	−1

For a directed connected graph with n nodes any $n-1$ rows of the incidence matrix forms a basis. However, the method of deriving the last row is not the same as for the undirected graph. To obtain the nth row add the first $n-1$ rows algebraically and change the sign of each term so obtained. The justification of this rule will be left to the reader.

2.7. The Circuit Matrix

The circuit matrix, C, provides a second interesting way of representing a graph. In this matrix a column is assigned to each arc of the graph and a row to each of a number of circuits of the graph. The number of circuits that must be considered will be discussed later in this section. Each circuit included in the matrix is represented as follows. If arc j belongs to circuit C_i, then c_{ij} (an element of the circuit matrix) takes the value one. Otherwise $c_{ij} = 0$.

Consider the graph in Fig. 2.24. Its three circuits are indicated by arrows and its circuit matrix is:

28 *Graphs and Networks*

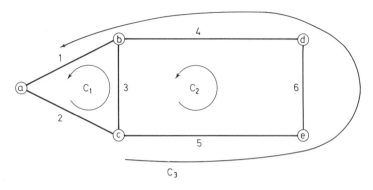

Fig. 2.24

$$\begin{array}{c|cccccc} & 1 & 2 & 3 & 4 & 5 & 6 \\ \hline C_1 & 1 & 1 & 1 & 0 & 0 & 0 \\ C_2 & 0 & 0 & 1 & 1 & 1 & 1 \\ C_3 & 1 & 1 & 0 & 1 & 1 & 1 \end{array}$$

When the graph is small it is not difficult to identify all the circuits and so construct the circuit matrix. However for a larger graph the process of identifying all the circuits is difficult and in the end unnecessary.

With the incidence matrix we found that any $n-1$ rows could be used to generate the nth row. With the circuit matrix it is also possible to find a basis from which all circuits of the graph can be generated. Generally the basis will be considerably smaller than the entire matrix. Most graphs can have different bases of circuits any one of which can be the *basic circuit matrix* for the graph.

To obtain a basic circuit matrix for a graph G consider H which is some tree of G. Now H contains no circuit but the addition to H of one arc of $\bar{H}$ creates one circuit (called a *fundamental circuit*) and one only. The matrix representing the set of circuits formed by adding each arc of $\bar{H}$ to H in turn forms a basic circuit matrix of G. There will be as many different basic circuit matrices for G as there are trees.

Any ring sum of basic circuits, $C_{i_1} \oplus C_{i_2} \oplus C_{i_3} \oplus \ldots C_{i_k}$ will be a circuit of G providing that for $j = 1, 2, 3 \ldots k$, the circuit $C_{i_{(j+1)}}$ has at least one arc in common with the circuit formed by the ring sum of the first j basic circuits. If all possible sums having this property are taken then all the circuits of G will be generated. Fig. 2.25 and the associated table illustrate this process.

It is easy to see that the set of fundamental circuits is independent, as each fundamental circuit contains at least one arc which is not contained in any other circuit of the set. To show that the set contains enough circuits to be a basis for the generation of all circuits of the

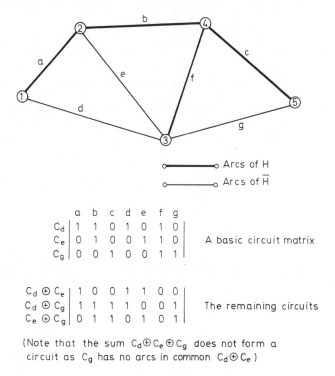

Fig. 2.25

graph is more difficult and requires certain concepts regarding vector spaces which are not covered in this book.[2]

The construction of a circuit matrix for a directed graph is complicated somewhat by the orientation of the arcs. When a tree is used as previously to define the basis we will obtain fundamental cycles and not fundamental circuits as there is no guarantee that all the arcs of a cycle so defined will be orientated in the same direction (some directed graphs contain no circuits, see Fig. 2.26). Secondly we must find a way of retaining a record of the orientation of the arcs in the matrix. Finally we must re-define our rule of addition for generating the complete set of circuits of the graph.

For a directed graph we will find a set of fundamental cycles by selecting a tree H and proceeding as previously. Each cycle is formed

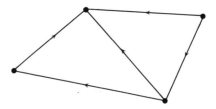

Fig. 2.26. A directed graph with no circuits

of arcs of H and one arc of $\bar{H}$. When the matrix is formed those arcs of a fundamental cycle which are orientated in the same direction as the arc of $\bar{H}$ will be assigned the matrix entry $+1$ and those orientated in the opposite direction will be assigned the entry -1. This procedure is illustrated in Fig. 2.27 and the associated table.

In generating new cycles we proceed in almost the same manner as before. However we must establish some arbitrary order of dominance among the basic cycles. This will permit the orientation of a new cycle formed by combination to be determined unambiguously (when several basic cycles are combined the resulting cycle is given the orientation of the dominant cycle in the combination). The ring sum ($\oplus$) is replaced by the *ring sum-difference* ($\pm$).

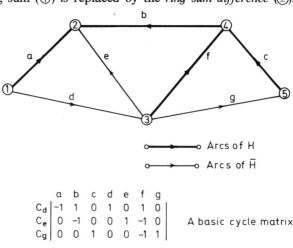

Fig. 2.27

This operation is defined as follows. When the ring sum-difference of two cycles C_i and C_j is taken (and C_i dominates C_j), the arcs common to the two cycles should be examined. If they have opposite signs in the two rows of digits the rows are added. If the signs are the same, the dominated row C_j is subtracted from the dominant row C_i. This has the effect of eliminating the common arcs as was done in the case of the non-directed graph and of ensuring that the arcs of the resulting cycle are given signs in accord with the orientation of the dominant cycle in the combination (Fig. 2.27).

2.8. The Co-circuit Matrix

The co-circuit matrix D is formed in much the same way as the circuit matrix. A column of the matrix is assigned to each arc of G, and each row of the matrix represents some co-circuit D_i. If arc j is a member of co-circuit i then the ij entry in the matrix is one. Otherwise the entry is zero. The co-circuit matrix for Fig. 2.24 is as follows:

	1	2	3	4	5	6
D_1	1	1	0	0	0	0
D_2	1	0	1	0	0	1
D_3	0	0	0	1	0	1
D_4	0	0	0	0	1	1

The matrix above does not contain all the co-circuits of the graph to which it refers. It is in fact a *basic co-circuit matrix* for that graph and from it all the co-circuits of that graph can be generated.

To obtain a basic co-circuit matrix for a graph G consider H, a tree of G, and $\bar{H}$ the corresponding co-tree. $\bar{H}$ contains no co-circuits but the addition to $\bar{H}$ of one arc of H will create a unique, so called fundamental co-circuit. The set of fundamental co-circuits which is found by the addition in turn of each arc of H to $\bar{H}$ forms a basis of co-circuits for the graph.

A basic co-circuit matrix for the graph of Fig. 2.25 is as follows. It corresponds to the indicated choice of tree and co-tree.

	a	b	c	d	e	f	g
D_a	1	0	0	1	0	0	0
D_b	0	1	0	1	1	0	0
D_c	0	0	1	0	0	0	1
D_f	0	0	0	1	1	1	1

The remaining co-circuits are generated in the same way as

circuits are generated from the basic circuit matrix. Other co-circuits so generated are:

$$\begin{array}{c|ccccccc} D_a \oplus D_b & 1 & 1 & 0 & 0 & 1 & 0 & 0 \\ D_b \oplus D_f & 0 & 1 & 0 & 0 & 0 & 1 & 1 \\ D_b \oplus D_c \oplus D_f & 0 & 1 & 1 & 0 & 0 & 1 & 0 \end{array}$$

For a directed graph a basic *co-cycle matrix* is formed (just as a basic co-circuit matrix is formed for the non-directed graph) by selecting a tree H and a corresponding co-tree $\bar{H}$. The arcs of H are sometimes called *fundamental arcs* and the co-cycles formed by the addition of a fundamental arc to $\bar{H}$ are called *fundamental co-cycles*.

In the matrix representation of a set of basic co-cycles a column is assigned to each arc of the graph and a row to each fundamental co-cycle. If arc i is a member of co-cycle j and is orientated in the same direction as the fundamental arc of that co-cycle, the entry ij takes the value $+1$. If it is in the opposite direction the value of the entry is -1. If the arc is not a member of the co-cycle the entry is zero.

A basic co-cycle matrix for the directed graph of Fig. 2.27 is shown in the table below. It corresponds to the tree shown in the Figure.

	a	b	c	d	e	f	g
D_a	1	0	0	1	0	0	0
D_b	0	1	0	-1	1	0	0
D_c	0	0	1	0	0	0	-1
D_f	0	0	0	-1	1	1	1

To generate the remaining co-cycles of the graph an order of dominance is established and co-cycles are combined using the ring sum-difference. For the example considered, let $D_a > D_b > D_c > D_f$. Some other co-cycles of the graph are:

	a	b	c	d	e	f	g
$D_a \pm D_b$	1	1	0	0	1	0	0
$D_a \pm D_f$	1	0	0	0	1	1	1
$D_b \pm D_f$	0	1	0	0	0	-1	-1
$D_c \pm D_f$	0	0	1	-1	1	1	0
$D_b \oplus D_c \pm D_f$	0	1	-1	0	0	-1	0
$D_a \pm D_c \pm D_f$	1	0	1	0	1	1	0

2.9. Relations Between Matrices

The three matrices which have been mentioned here are not the only ones which can be constructed from a graph, but they are the most important. Others are described in reference 3.

These matrices are linked by certain mathematical relationships. We find that $EC' = 0$, where C' is the transpose of the circuit matrix. This can be verified by considering the incidence and circuit matrices from the graph of Fig. 2.25:

$$\begin{vmatrix} 1 & 0 & 0 & 1 & 0 & 0 & 0 \\ 1 & 1 & 0 & 0 & 1 & 0 & 0 \\ 0 & 0 & 0 & 1 & 1 & 1 & 1 \\ 0 & 1 & 1 & 0 & 0 & 1 & 0 \\ 0 & 0 & 1 & 0 & 0 & 0 & 1 \end{vmatrix} \begin{vmatrix} 1 & 0 & 0 \\ 1 & 1 & 0 \\ 0 & 0 & 1 \\ 1 & 0 & 0 \\ 0 & 1 & 0 \\ 1 & 1 & 1 \\ 0 & 0 & 1 \end{vmatrix} = \begin{vmatrix} 0 & 0 & 0 \\ 0 & 0 & 0 \\ 0 & 0 & 0 \\ 0 & 0 & 0 \\ 0 & 0 & 0 \end{vmatrix}$$

(remember that addition is done modulo 2 and that E and C' must be arranged so that the columns of E and the rows of C' are in the same order of arcs).

It is not difficult to see why this relation should hold. Each cofficient in the product matrix is the scalar product of a row of E and a column of C'. Each row of E shows which arcs are incident to a given node. No circuit will contain more than two arcs from any row of E as this would imply that some node is adjacent to more than two arcs of the cycle which by definition is not possible. Therefore a column of C' representing a circuit will have either two non-zero entries in the same position as the non-zero entries of a row of A or none at all. The scalar product then is either 0 or 0 (mod 2).

It is also true that $CD' = 0$, where D' is the transpose of the basic co-circuit matrix. This relation can be verified using the matrices derived from the graph of Fig. 2.25:

$$\begin{vmatrix} 1 & 1 & 0 & 1 & 0 & 1 & 0 \\ 0 & 1 & 0 & 0 & 1 & 1 & 0 \\ 0 & 0 & 1 & 0 & 0 & 1 & 1 \end{vmatrix} \begin{vmatrix} 1 & 0 & 0 & 0 \\ 0 & 1 & 0 & 0 \\ 0 & 0 & 1 & 0 \\ 1 & 1 & 0 & 1 \\ 0 & 1 & 0 & 1 \\ 0 & 0 & 0 & 1 \\ 0 & 0 & 1 & 1 \end{vmatrix} = \begin{vmatrix} 0 & 0 & 0 & 0 \\ 0 & 0 & 0 & 0 \\ 0 & 0 & 0 & 0 \end{vmatrix}$$

This relationship must hold because a circuit and co-circuit must have an even number of arcs in common. Suppose that this were not so and that circuit C_i and co-circuit D_j have only one arc in common (remember that D_j divides the nodes of the graph into two sets S_1 and S_2 and that the arcs of D_j are those arcs which join S_1 to S_2). Suppose that we start tracing C_i through the graph with a pencil marking its arcs in heavy lines. If we start at some node of S_1 we will at some point cross into S_2 using the one arc which C_i and D_j

have in common. We are now trapped in S_2 as there is no other arc of C_i which can be used to cross back into S_1. Thus C_i is not a circuit which contradicts the original statement. This line of reasoning can be extended to show that a higher odd number of common arcs is also impossible to obtain. Therefore the scalar product of a row of C and a column of D' must be an even number which is zero, modulo 2.

Matrices E and C and matrices C and D are said to be *orthogonal*, just as two vectors are said to be orthogonal if their scalar product is zero.

2.10. Matroids and Graphs [9,5]

The symbols used in this definition are those of Set Theory. They are read as follows:

$\in$ — 'is an element of', used to show that an element is a member of some set.

$\subset$ — 'is contained in', indicates that the set preceding the symbol is a subset of the set following it.

$\cup$ — 'union', $A \cup B$ is the set of elements belonging to either A or B or both.

$\cap$ — 'intersection', $A \cap B$ is the set of elements belonging to both A and B.

A *matroid* $M(L,C)$ is a mathematical structure consisting of a

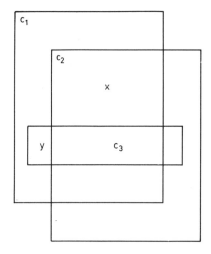

Fig. 2.28

finite set L of elements and a collection C of non empty subsets of L satisfying the following two conditions:
(1) No subset in the collection C contains another subset properly.
(2) If C_1 and C_2 are subsets of the collection C, and x and y are elements of L such that $x \in C_1 \cap C_2$ and $y \in (C_1 - C_2)$ then there exists a subset C_3 such that $y \in C_3$, $x \notin C_3$, and $C_3 \subset (C_1 \cup C_2)$. This second condition is illustrated in the Venn Diagram of Fig. 2.28.

Example 1: Let the set L consist of five objects (a, b, c, d, e). Let us consider the collection C of all subsets of L having exactly four objects. We can represent this collection by a matrix of zeros and ones called the representative matrix of the matroid. However note that not all matroids correspond to matrices. The rows represent the subsets C_i and the columns correspond to the objects of L. If the ij^{th} entry is 1, the jth object is a member of the ith subset. Otherwise the entry would take the value 0. The structure described above is represented as follows:

	a	b	c	d	e
C_1	0	1	1	1	1
C_2	1	0	1	1	1
C_3	1	1	0	1	1
C_4	1	1	1	0	1
C_5	1	1	1	1	0

Example 2: Considering all the subsets D_i containing exactly three members of L we obtain:

	a	b	c	d	e
D_1	0	0	1	1	1
D_2	0	1	0	1	1
D_3	0	1	1	0	1
D_4	0	1	1	1	0
D_5	1	0	0	1	1
D_6	1	0	1	0	1
D_7	1	0	1	1	0
D_8	1	1	0	0	1
D_9	1	1	0	1	0
D_{10}	1	1	1	0	0

It is evident that these two structures satisfy condition (1) in the definition of a matroid. Verifying condition (2) for all possible pairs of subsets with common elements is tedious but consider the following case:
let the C_1 of condition (2) be C_1 of example 1
let the C_2 of condition (2) be C_2 of example 1

let the C_3 of condition (2) be C_3 of example 1
let the x of condition (2) be the element c of L in example 1
let the y of condition (2) be the element b of L in example 1
With these specifications we obtain:

$$c \in C_1 \cap C_2$$
$$b \in (C_1 - C_2)$$
$$b \in C_3$$
$$c \notin C_3$$
$$C_3 \subset (C_1 \cup C_2)$$

The reader will not find it difficult to verify other cases in these examples.

2.11. Matroids, Graphs and Matrices

It can be shown that the complete circuit matrix and the complete co-circuit matrix of a graph both satisfy the definition of a matroid where L is interpreted as the set of arcs of the graph, C as the set of circuits and D as the set of co-circuits. In what follows we will refer to these matrices as the *circuit matroid* and the *co-circuit matroid* of the graph. The circuit and co-circuit matroids of the graph of Fig. 2.24 are:

```
        1 2 3 4 5 6              1 2 3 4 5 6
   C₁  |1 1 1 0 0 0|        D₁  |1 1 0 0 0 0|
   C₂  |0 0 1 1 1 1|        D₂  |1 0 1 0 0 1|
   C₃  |1 1 0 1 1 1|        D₃  |0 0 0 1 0 1|
                            D₄  |0 0 0 0 1 1|
```

However not every matroid corresponds to a graph. An attempt to construct a graph using the matroids of Example 1 and Example 2 as a circuit or co-circuit matroid will quickly fail.

2.12. The Dual Matroid

Every matroid has a dual. The definition of the dual matroid in general terms is quite lengthy although for those matroids which correspond to matrices the definition is simple. If C is the representative matrix of the matroid $M(L,C)$ and D is the representative matrix of the matroid $M(L,D)$, then $M(L,C)$ and $M(L,D)$ are duals if (in matrix notation):

$$CD' = DC' = 0$$

That is to say C and D are orthogonal. The two matroids shown in Example 1 and Example 2 are duals to each other as are the two matroids shown in the previous subsection.

2.13. Planar Graphs and Matroids

Let G be a planar graph and G' its dual. Let the circuit matroid of G be $M(L,C)$ and let the co-circuit matroid be $M(L,D)$. It can be shown that $M(L,C)$, the circuit matroid of G serves also as the

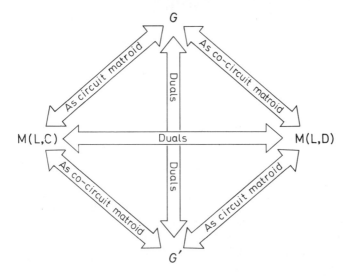

Fig. 2.29. For planar graphs

co-circuit matroid of G'. Similarly $M(L,D)$ which is the co-circuit matroid of G serves also as the circuit matroid of G'. These relationships are illustrated in Fig. 2.29.

2.14. Non-planar Graphs and Matroids

If the graph G is non-planar the situation is changed because the dual graph G' does not exist. Of course the circuit matroid $M(L,C)$ still exists but in this case it cannot serve as the co-circuit matroid of any graph. Similarly $M(L,D)$ the co-circuit matroid of G cannot serve as the circuit matroid of any graph. Fig. 2.30 illustrates the relationships between a non-planar graph and its matroids.

A matroid which can serve as the co-circuit matroid of a graph is said to be *graphic*. A matroid which can serve as the circuit matroid of some graph is said to be *co-graphic*. If a matroid is both graphic and co-graphic then by definition the corresponding graphs are

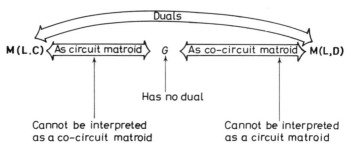

Fig. 2.30. *For non planar graphs*

planar and duals to each other. At the opposite end of the situation is the matroid which corresponds to no graph and is therefore neither graphic nor co-graphic.

2.15. Constructing a Graph from a Matrix

Electrical engineers are often faced with the problem of constructing an electrical network that will satisfy certain requirements such as which elements must be in series, (that is to say belong to the same cycle) or which must be in parallel (belong to the same co-cycle). From these requirements a matrix will be constructed which the engineer hopes will be the cycle or co-cycle matroid of some graph so that the required network may be constructed. The following situations are possible:

(1) The matrix is not a matroid. In this case one may state immediately that the desired network cannot be constructed.
(2) The matrix is a matroid but the required network cannot be constructed. This will happen if the elements of the matroid represent circuits of the desired network and the matroid is found to be graphic but not co-graphic.
(3) The matrix is a matroid and the required network can be constructed from it.

For a small matrix it may be easy to determine which of these conditions holds. For a large problem even if conditions in (3) hold, the construction of the graph would be very difficult without an organised approach. W. T. Tutte[7] has described an algorithm which

EXERCISES

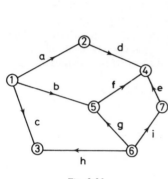

Fig. 2.31

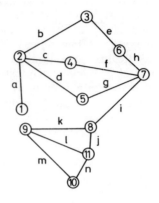

Fig. 2.32

1. In Fig. 2.31 is bgi a path?
2. In Fig. 2.31 is $bfda$ a circuit?
3. (a) Find a spanning tree in the graph of Fig. 2.32 and verify for yourself that it satisfies the six definitions of a spanning tree which are given in the text.
   (b) Is there a spanning tree not containing arc i in Fig. 2.32? Given your answer, state what type of graph this is.
   (5, 6)? Is either of these two co-cycles a co-circuit?
5. In Fig. 2.32 is $efkj$ a co-tree? Find a co-tree not containing arc e.
6. Find as many non-planar graphs having six nodes as you can. Do some of them have a common characteristic?
7. Find all the cuts between node 1 and node 7 of Fig. 2.31. Use the dual graph to help you to find them.
8. Find the incidence matrix, the circuit matrix and the co-circuit matrix of the graph in Fig. 2.8. Verify for yourself that $EC' = 0$ and $CD' = 0$.
9. Assume the matrix shown below to be the cycle matrix of some graph. Construct the graph that it represents.

$$\begin{array}{c|cccccccc} C_1 & 1 & 1 & 0 & 0 & 1 & 0 & 0 & 0 \\ C_2 & 0 & 0 & 0 & 1 & 1 & 1 & 0 & 0 \\ C_3 & 0 & 0 & 0 & 0 & 0 & 1 & 1 & 1 \\ C_4 & 0 & 1 & 1 & 0 & 0 & 0 & 1 & 0 \end{array}$$

To what tree does this cycle matrix correspond?

10. In exercise 9 verify that the ring sum of any pair of the basic cycles is also a cycle.
11. Construct a co-cycle matrix for the graph of exercise 9.
12. In the matrix of exercise 9 let the entries of columns 1, 3, 4 and 8 remain positive and assign positive and negative signs to the remaining entries in some arbitrary fashion. Repeat exercises 9 through 11 for the directed graph this matrix now represents. (Let the order of dominance be $C_1 > C_2 > C_3 > C_4$.)
13. Revise the definitions of *arc adjacency, degree, directed graph* and *chain* given in the text to allow for the existence of self-loops. Test your definitions by attempting to draw graphs having structures not covered in the definitions.
14. Partition the arcs of the graph of Fig. 2.32 into disjoint subsets each of which is either a chain or a cycle (such a partition is called a *covering*). Attempt to find the smallest number of subsets with which this can be done (this is the *minimal covering*).

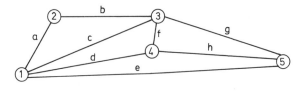

Fig. 2.33

15. .Find the minimal covering for the graph of Fig. 2.33.
16. Show that the intersection of a cycle with a co-cycle must be an even number of arcs.
17. Show that the translations from graph to circuit matrix and back to the graph are unique.
18. *The Assignment Problem*[8]: A, B, C, D and E are men who are to be assigned to jobs 1, 2, 3, 4 and 5. A can do any of the jobs, B can do all but job 3, C can do only 1 and 4, D can do 2, 4 and 5, and E can do any job. Model this situation by means of a graph. Under what conditions will no solution to this type of problem be available?
19. *Choosing Matches*: A pile of n matches and a pile of $n+1$ matches are placed on a table. Two players, white and black, take turns choosing one match from either pile or one match from each with white making first choice. The players continue to pick up matches until none are left. The player picking up the last match wins. The moves in this game can be illustrated using a graph in which the nodes are given a label of the form (x, y) indicating that there remain x matches in one pile and y in the other, (note that the situation (x, y) is the same as the situation (y, x)). For a small value of n draw the

graph showing all possible moves and try to deduce from it the conditions under which each player may win.

20. *Tournaments:* In a round-robin tournament each player must meet each other player in one match. If n is the number of players (and if n is even) each player will play $n-1$ matches. With the aid of a graph representing all possible pairings draw up a table of matches in a six person round-robin tournament.

21. *The Decanting Problem:* An eight litre wine jug is full. Given a five litre jug and a three litre jug (which are empty), divide the wine into two portions of four litres without using any other measuring instrument. Find your solution by using a graph in which each node has a label showing how much wine is contained in the five and three litre jugs.

22. *The Ferryman's Problem:* A ferryman must carry a wolf, a goat and a giant cabbage across a river in a boat so small that he can take only one item per trip. Furthermore he cannot leave the wolf and goat alone together, nor can the cabbage and goat be left unattended together. Use a graph (having three nodes on each river bank) to show how all three passengers may be ferried across.

23. *The Analysis of Interconnected Decision Areas*[6]: In such fields as engineering and architectural design it is often required to choose from among a number of options in each of a number of areas. However not every option in one area is compatible with every option in another area. Let us consider the following simple example of furnishing an office:

It is required to choose from among three chairs (CH_1, CH_2, CH_3), three desks (D_1, D_2, D_3), two bookcases (B_1, B_2), three colours of curtains (C_1, C_2, C_3), and three types of carpets (CA_1, CA_2, CA_3). Because of clashes in colour and style, the following incompatibilities exist:

CH_2 does not match with CA_2, D_1, C_1, and C_2
CH_3 ,, ,, ,, ,, D_3
C_1 ,, ,, ,, ,, D_3 and CA_1
C_2 ,, ,, ,, ,, CA_1
B_1 ,, ,, ,, ,, D_1 and D_2
B_2 ,, ,, ,, ,, D_3

Represent this situation by means of a graph and find at least one compatible set of furnishings.

ANSWERS TO SELECTED EXERCISES

9.

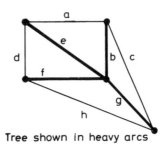

Tree shown in heavy arcs

Fig. 2.34

14. One covering is: $(abehi)$, $(dcfg)$, (kjl), (mn).

15. This graph is said to be *unicursal* because the minimal covering is one chain: $(dabgecfh)$.

18. There is a solution to the problem of assigning n men to n jobs if any subset of k men ($k = 1, 2, \ldots n$), is collectively able to do at least k jobs. (For example C and D are collectively capable of doing four jobs.) This is Philip Hall's theorem.[8]

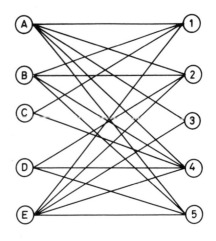

Fig. 2.35

19. Assume that Player A chooses first (Fig. 2.36).

Graphs 43

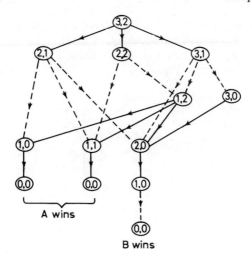

Fig. 2.36

The graph above represents the case where $n = 2$. Player A's possible choices are shown by continuous lines and Player B's choices by broken lines.

20. The matches in the first round are indicated by arcs marked 1 and so on (Fig. 2.37).

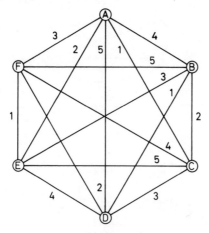

Fig. 2.37

21. If one can bring about the situation where together, the five and three litre jugs hold four litres of wine, the problem is solved. There-

fore a solution consists in finding a feasible path from the node (0, 0) to any of the four circled nodes ((4, 0), (3, 1), (2, 2), (1, 3)). One such path is shown. Is it the one representing a solution in the least number of pourings? (Fig. 2.38).

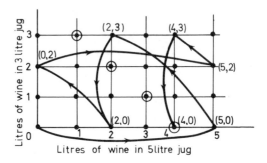

Fig. 2.38

22. The goat is ferried across first. Broken lines represent empty crossings. Arcs joining nodes on the same bank indicate that one passenger was dropped off and the other picked up (Fig. 2.39).

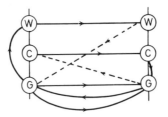

Fig. 2.39

23. Arcs represent incompatibilities. (B_1, D_3, CH_1, C_3) is a compatible combination (Fig. 2.40).

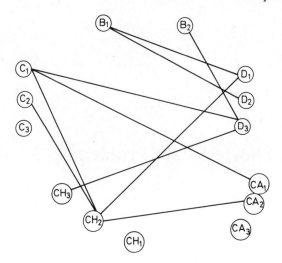

Fig. 2.40

REFERENCES

1. BERGE, C. et GHOUILA-HOURI, A., *Programmes, Jeux, et Réseaux de Transport*, Dunod, Paris (1962).
2. BUSACKER, R. G. and SAATY, T. L., *Finite Graphs and Networks*, McGraw-Hill, New York (1965).
3. HARARY, F., NORMAN, R. and CARTWRIGHT, D., *Structural Models: Introduction to the Theory of Directed Graphs*, Wiley, New York (1965).
4. MAXWELL, L. M. and REED, M. B., 'Graphs', Published Lecture Notes, Colorado State University (1967).
5. MINTY, G. J., 'On the Axiomatic Foundations of the Theories of Directed Linear Graphs, Electrical Networks, and Network Programming', *J. Math. and Mech.*, **15** (1966).
6. O'FLAHERTY, T. B., 'The Analysis of Interconnected Decision Areas', *Bulletin of the Canadian Operational Research Society* (May 1969).
7. TUTTE, W. T., 'An Algorithm for Determining Whether a Given Binary Matroid is Graphic', *Proc. Amer. Math. Soc.*, **11** (1960).
8. VAJDA, S., *Mathematical Programming*, Addison-Wesley (1961).
9. WHITNEY, H., 'On the Abstract Properties of Linear Dependence', *Amer. J. Math.*, **57** (1935).

CHAPTER 3

Shortest Path Problems

3.1. Shortest Path Algorithms

Fig. 3.1 shows a two-terminal network having a set of arcs A and a set of nodes N (consisting of a source S, a sink T, and three other nodes). To each arc joining a pair of nodes (i, j) is attached a number $a(ij)$ which is the *length* of the arc. The problem is to find the shortest path from S to T. Several algorithms have been proposed[6]. The first which will be described here has been cited by Dreyfus[2] (who credits it to Dijkstra)[1] as being the most efficient.

3.2. The Dijkstra Algorithm

In this algorithm each node is assigned a two part *label* which may be either *permanent* or *tentative*. Suppose that a node p were assigned a permanent label of the form $[q, \pi(p)]$. This label would indicate that the shortest path from S to p is of length $\pi(p)$ and that the node previous to p on this path is q. If the label were tentative, the two parts would have the same meaning but would refer only to the best path so far found.

Step 1 Assign to the source the permanent label $(-, 0)$ and to all other nodes the tentative label $(-, \infty)$. In the case of the source this indicates that the shortest path from S to S is of length zero and contains no intervening nodes. For all other nodes the label indicates that no paths are yet known and so no preceding node can be

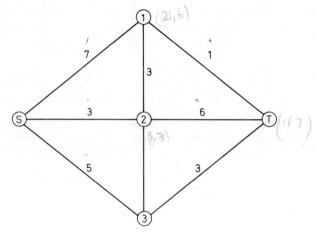

Fig. 3.1

specified. By convention the second part of the label is designated infinity. Let $k = S$ and proceed directly to Step 3.

Step 2 Let S_t be the set of nodes bearing tentative labels. Find node k such that $\pi(k) = \min_{x \in S_t}[\pi(x)]$. Declare the label of node k to be permanent. If $k = T$, terminate, as the shortest path from S to T has been found. Otherwise proceed to Step 3.

Step 3 Consider each node y which is directly connected to k and which bears a tentative label. If $\pi(k) + a(ky) < \pi(y)$ a new tentative label is assigned to y. The first part of the label is changed to k and $\pi(y)$ takes the new value $\pi(k) + a(ky)$. If $\pi(k) + a(ky) \geqslant \pi(y)$ no change is made. When all such nodes y have been considered return to Step 2.

When the algorithm has terminated, the shortest path from S to T can be found by using the first part of the labels to trace backwards from T to S.

The proof that this method yields the shortest path is inductive. Consider the situation following the first application of Step 2. Some node k has been assigned a permanent label, and so now two nodes (S and k) bear permanent labels while the remainder are tentative. The label of k asserts that the shortest path from S to k is the direct connection between these nodes. Were this not so the shortest path would pass through at least one other node before reaching k. But since $\pi(k) = \min_{x \in S_t}[\pi(x)]$, all other nodes are farther from S than k itself and no shorter path can exist.

Having established the validity of the first application of the routine the situation at any following step is as follows. In general the nodes are divided into two sets, S_p (those bearing permanent labels) and S_t (those bearing tentative labels), where $S_p \cup S_t = N$, and $S_p \cap S_t = \phi$. The labels of the nodes in S_p indicate the shortest paths from S to these nodes, whereas a non-infinite label attached to a node of the set S_t indicates merely the shortest path to this node containing only nodes of S_p. Now $k \in S_t$ and $\pi(k) = \min_{x \in S_t}[\pi(x)]$. Node k can therefore be transferred to S_p because if some path of length less than $\pi(k)$ exists, it is one which contains at least one other node of S_t. But all such nodes are farther from S than k itself and so $\pi(k)$ is the true minimum distance from S to k.

Solving a small sample problem using this method demonstrates how the set S_p grows until it includes the terminal node T. Consider the network shown in Fig. 3.1. To aid solution a tableau is formed from the distance matrix of the network. Both tentative and permanent labels are recorded along the top of the matrix and permanent labels only down the left hand side. Permanent labels are indicated by a superscript 'p', while the node designated as 'k' at each iteration bears this letter as a subscript to its label. The solution to the problem is as follows (Fig. 3.2).

Note that in this example the algorithm has found the shortest path from S to all other nodes because by the time the label on T is declared permanent all other labels have also been declared permanent. This will not always occur but the shortest paths to all nodes can always be found by not terminating until all labels are permanent.

3.3. The Ford Algorithm

The Dijkstra algorithm depends for its validity on the condition that all $a(ij)$ be greater than or equal to zero. If this is not so the procedure for declaring labels permanent can give incorrect results. For example a network could exist in which some arcs represent stages of a journey which incur a cost (the price of a ticket) and others which return a profit (the traveller works to pay his passage). The objective of the traveller is to complete his journey from S to T at minimum cost and if possible at negative cost (profit). The algorithm due to Ford[4] will solve this problem providing there is no circuit in which the sum of the arc lengths is negative. If there is such a circuit the algorithm will cycle around it reducing cost at each cycle without terminating.

Step 1 Assign the label $(-, 0)$ to the source node S, and the label $(-, \infty)$ to all other nodes.

	$(-,0)_K$	$(-,\infty)$	$(-,\infty)$	$(-,\infty)$	$(-,\infty)$
S		7	3	5	
1	7		3		1
2	3	3			6
3	5				3
T		1	6	3	

First labelling

	$(-,0)^P$	$(S,7)$	$(S,3)_K$	$(S,5)$	$(-,\infty)$
$(-,0)^P$ S		7	3	5	
1	7		3		1
2	3	3			6
3	5				3
T		1	6	3	

Second labelling

	$(-,0)^P$	$((2),6)$	$(S,3)^P$	$(S,5)_K$	$((2),9)$
	S	1	2	3	T
$(-,0)^P$ S		7	3	5	
1	7		3		1
$(S,3)^P$ 2	3	3			6
3	5				3
T		1	6	3	

Third labelling

	$(-,0)^P$	$((2),6)_K$	$(S,3)^P$	$(S,5)^P$	$((3),8)$
	S	1	2	3	T
$(-,0)^P$ S		7	3	5	
1	7		3		1
$(S,3)^P$ 2	3	3			6
$(S,5)^P$ 3	5				3
T		1	6	3	

Fourth labelling

	$(-,0)^P$	$((2)6)^P$	$(S,3)^P$	$(S,5)^P$	$((1)7)_K$
	S	1	2	3	T
$(-,0)^P$ S		7	3	5	
$((2),6)^P$ 1	7		3		1
$(S,3)^P$ 2	3	3			6
$(S,5)^P$ 3	5				3
T		1	6	3	

Final labelling

Fig. 3.2

Step 2 Search for an arc (*ij*) such that $\pi(i)+a(ij) < \pi(j)$. If no such arc exists terminate. Otherwise proceed to Step 3.

Step 3 Change to *i* the first part of the label of node *j*. Change the second part of the label to $\pi(i)+a(ij)$. Return to Step 2.

Upon termination the shortest path from *S* to *T* is found by tracing the labels back from *T* as for the previous algorithm. To be rigorous, it must be proved that the algorithm terminates without cycling and identifies a path which is in fact the shortest. Heuristically though, it is not difficult to see that the path from *S* to *T* indicated upon termination is optimal. Suppose that after the final execution of Step 2 the sink is labelled $[j, \pi(T)]$ but that some path $(S, n_1, n_2, n_3, \ldots n_m, T)$ of length $\pi' < \pi(T)$ exists. This implies that *T* can be labelled (n_m, π') either immediately, or after changing one or more labels on other nodes of the path. However the condition for termination of the algorithm is that it is not possible to change any more labels. This contradicts the statement that Step 2 can no longer be executed and therefore no path shorter than $\pi(T)$ can exist.

A solution tableau similar to that used for the Dijkstra algorithm can be used to solve small problems. The labels are recorded both along the top and down the left hand side. After the initial labels are fixed a 'pass' is made through the arcs of the network searching for those which satisfy the condition of Step 2. New labels are affixed where possible and the process is repeated. The solution to the problem of Fig. 3.1 is illustrated (Fig. 3.3).

3.4. The Floyd Algorithm

If the shortest path between all pairs of nodes is required the Dijkstra or Ford algorithms can be applied *N* times using each node in turn as source node. However other methods exist for finding the shortest distance matrix. Floyd's algorithm[3] uses the $N \times N$ matrix of direct distances between nodes as input. Nodes not joined by an arcs are assigned a direct distance of infinity and nodes are (by convention) joined to themselves by an arc of length zero. Self-loops are disregarded.

Step 1 Form the $N \times N$ distance matrix and call it D^1. Let $k = 0$.

Step 2 Increase the value of *k* by one. If $k = N+1$, terminate; otherwise go to Step 3.

Step 3 Let $D^{(k+1)}$ be the matrix having the elements:

$$d_{ij}^{(k+1)} = \min[d_{ij}^{(k)}, d_{ik}^{(k)} + d_{kj}^{(k)}]$$

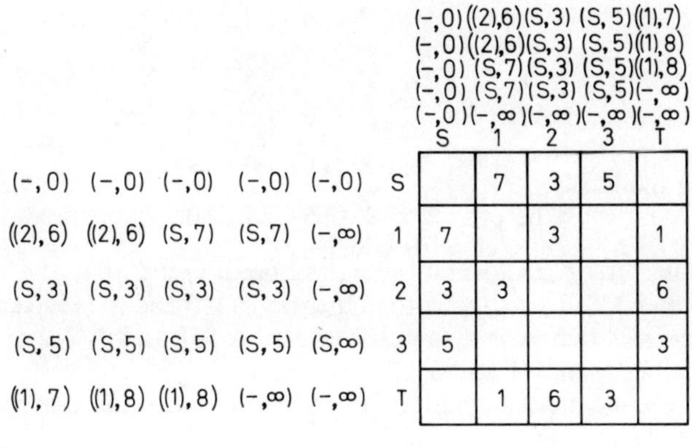

$$\begin{array}{c}
(-,0)\;((2),6)\;(S,3)\;(S,5)\;((1),7)\\
(-,0)\;((2),6)\;(S,3)\;(S,5)\;((1),8)\\
(-,0)\;\;(S,7)\;(S,3)\;(S,5)\;((1),8)\\
(-,0)\;\;(S,7)\;(S,3)\;(S,5)\;(-,\infty)\\
(-,0)\;(-,\infty)(-,\infty)(-,\infty)(-,\infty)\\
\;\;\;\;S\;\;\;\;\;\;\;1\;\;\;\;\;\;\;2\;\;\;\;\;\;\;3\;\;\;\;\;\;\;T
\end{array}$$

						S	1	2	3	T
(-,0)	(-,0)	(-,0)	(-,0)	(-,0)	S		7	3	5	
((2),6)	((2),6)	(S,7)	(S,7)	(-,∞)	1	7		3		1
(S,3)	(S,3)	(S,3)	(S,3)	(-,∞)	2	3	3			6
(S,5)	(S,5)	(S,5)	(S,5)	(S,∞)	3	5				3
((1),7)	((1),8)	((1),8)	(-,∞)	(-,∞)	T		1	6	3	

Fig. 3.3

Find the values of all the elements of this matrix and return to Step 2.

This algorithm produces a sequence of N matrices, the last of which is the matrix of shortest distances. At the first iteration the direct distance from node i to node j (d_{ij}) is replaced by the sum $(d_{i1} + d_{1j})$ if this quantity is less than d_{ij}. At the kth iteration the distances shown between node pairs will, in general, be for paths of more than one arc. At this point the algorithm searches for node pairs ij such that the shortest known distance between i and j is greater than the sum of the shortest known distances between pairs ik and kj. When such node pairs are found the new shorter distances are recorded.

For the network of Fig. 3.1 the direct distance matrix is:

$$D^1 = \begin{vmatrix} 0 & 7 & 3 & 5 & \infty \\ 7 & 0 & 3 & \infty & 1 \\ 3 & 3 & 0 & \infty & 6 \\ 5 & \infty & \infty & 0 & 3 \\ \infty & 1 & 6 & 3 & 0 \end{vmatrix}$$

At the first iteration we interpose node S between other pairs of nodes where this will give a shorter path than the direct distance. In the first transformed matrix shown below, those elements which have been changed are circled.

52 Graphs and Networks

$$D^2 = \begin{array}{c} S \\ 1 \\ 2 \\ 3 \\ T \end{array} \begin{array}{|ccccc|} 0 & 7 & 3 & 5 & \infty \\ 7 & 0 & 3 & \text{\textcircled{12}} & 1 \\ 3 & 3 & 0 & \text{\textcircled{8}} & 6 \\ 5 & \text{\textcircled{12}} & \text{\textcircled{8}} & 0 & 3 \\ \infty & 1 & 6 & 3 & 0 \end{array}$$

In this first transformation it has been found that the routes 1–S–3 and 2–S–3 are shorter than the routes 1–3 and 2–3 respectively. Because this matrix is symmetric routes 3–S–1 and 3–S–2 are also used rather than 3–1 and 3–2.

The second transformation produces the following matrix:

$$D^3 = \begin{array}{c} S \\ 1 \\ 2 \\ 3 \\ T \end{array} \begin{array}{|ccccc|} 0 & 7 & \text{\textcircled{3}} & 5 & \text{\textcircled{8}} \\ 7 & 0 & 3 & 12 & 1 \\ 3 & 3 & 0 & 8 & \text{\textcircled{4}} \\ 5 & 12 & 8 & 0 & 3 \\ \text{\textcircled{8}} & 1 & \text{\textcircled{4}} & 3 & 0 \end{array}$$

Here S–T has been replaced by S–1–T and 2–T by 2–1–T.

The third transformation gives:

$$D^4 = \begin{array}{|ccccc|} 0 & \text{\textcircled{6}} & 3 & 5 & \text{\textcircled{7}} \\ \text{\textcircled{6}} & 0 & 3 & \text{\textcircled{11}} & 1 \\ 3 & 3 & 0 & 8 & 4 \\ 5 & \text{\textcircled{11}} & 8 & 0 & 3 \\ \text{\textcircled{7}} & 1 & 4 & 3 & 0 \end{array}$$

In this transformation S–1 was replaced by S–2–1 and S–1–T is therefore replaced by S–2–1–T. The route 1–S–3 has also been replaced by 1–2–S–3.

The remaining two matrices are:

$$D^5 = \begin{array}{|ccccc|} 0 & 6 & 3 & 5 & 7 \\ 6 & 0 & 3 & 11 & 1 \\ 3 & 3 & 0 & 8 & 4 \\ 5 & 11 & 8 & 0 & 3 \\ 7 & 1 & 4 & 3 & 0 \end{array} \qquad D^6 = \begin{array}{|ccccc|} 0 & 6 & 3 & 5 & 7 \\ 6 & 0 & 3 & \text{\textcircled{4}} & 1 \\ 3 & 3 & 0 & \text{\textcircled{7}} & 4 \\ 5 & \text{\textcircled{4}} & \text{\textcircled{7}} & 0 & 3 \\ 7 & 1 & 4 & 3 & 0 \end{array}$$

In the final matrix 1–2–S–3 has been replaced by 1–T–3 and 2–S–3 has been replaced by 2–1–T–3.

Murchland has shown that this algorithm finds the shortest distance matrix correctly provided that the network to which it is applied has no circuits of negative length.

3.5. The Pollack Algorithm

Having found the shortest path from source to sink we may wish to find the second shortest path. Pollack,[8] in a simple method for doing this makes use of the fact that the second shortest path differs by at least one arc (or more) from the shortest path and with respect to this condition is the shortest path through the network.

Step 1 Find the shortest path through the network. Where there are m arcs in the shortest path let them be numbered from 1 to m. Let $i = 1$.

Step 2 The length of arc i of the shortest path is d_i. Temporarily, let $d_i = \infty$. Find and record the shortest path through this modified network.

Step 3 Restore d_i to its original value. Increase the value of i by one. If the new $i = m+1$ go to Step 4. Otherwise return to Step 2.

Step 4 The shortest of the paths found in Step 2 is the second shortest path. Terminate.

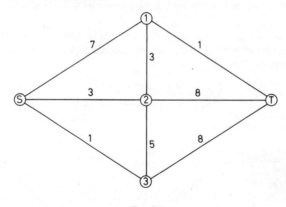

Fig. 3.4

The application of this algorithm to the network of Fig. 3.4 gives the following results:
at Step 1 the shortest path is S–2–1–T (length 7);
at the first iteration, path S–1–T (length 8);

at the second iteration, path S–1–T (length 8);
at the third iteration, path S–3–T (length 9).
We conclude that the second shortest path is S–1–T. The reader will have noticed that the same path was found at the first and second iteration. The algorithm has no means of avoiding the re-discovery of a known path.

3.6. The Hoffman and Pavley Algorithm

The algorithm of Hoffman and Pavley[5] is faster as it only involves one application of the shortest path algorithm, the remaining calculations being additions, subtractions and comparisons.

Step 1 Using T as the *origin* find the shortest paths from T to all other nodes using the Ford or Dijkstra algorithms. The label $\pi(i)$ therefore represents the shortest distance from i to T. Let the number of nodes on the shortest route from S to T (inclusive) be m. We will refer to these nodes by the index i and re-number them from 1 to m. Let $i = 1$.

Step 2 (a) Suppose that node i of the shortest route is connected to r other nodes of the network. Consider them to belong to a set J_i. Let j be some node of the set J_i.

(b) Consider the path which follows the shortest route as far as node i, deviates to node j, and then takes the shortest path from j to T. Record this path and its length which is $\{[\pi(s) - \pi(i)] + d_{ij} + \pi(j)\}$.

(c) If all nodes of J_i have been considered in Step 2b go to Step 2d. Otherwise let j be some node of J_i which has not yet been considered and return to Step 2b.

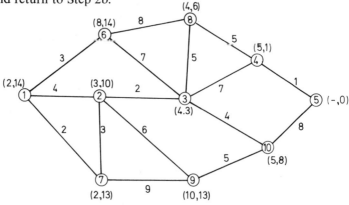

Fig. 3.5

(d) Increase the value of i by one. If $i \leqslant m$ return to Step 2a. Otherwise terminate, as the best path found in Step 2b is the second best path.

It will be noted that there is nothing to prevent the second best path found by this method from containing a loop. To prevent this it is necessary to specify that no path containing a loop be recorded in Step 2b.

The calculations done to find the second shortest path through the network of Fig. 3.5 are shown in the table below. The shortest path in the network was found using the Dijkstra algorithm. The permanent labels are shown in the Figure.

Node i	Sets J_i	$\{[\pi(1)-\pi(i)]+d_{ij}+\pi(j)\}$
1	6	17
	7	15
2	1	22 (loop)
	9	23
	7	20 (loop)
3	2	18 (loop)
	6	27
	8	17
	10	18
4	3	28 (loop)
	8	24 (loop)

The algorithm shows that the second shortest path is 1–7–2–3–4–5 and has a length of 15. Note that five of the paths found contain loops.

To show that the Hoffman and Pavley algorithm correctly finds the second shortest path it is useful to define that path as follows: it is the path which differs from the shortest path by at least one node (possibly more) and is the shortest path that respects this constraint. In differing from the shortest path it can bypass or replace nodes of the shortest path or form loops by adding one or more nodes to the set in the shortest path. Such a path will follow the shortest path for a number of nodes (which may be only one, the source), deviate to some node j, and then proceed to T by the shortest available route. By deviating to j we have insured that the path will be of length greater than or equal to that of the shortest path (there may be more than one path of optimal length). By then proceeding from j to T by the most direct route we ensure that the path is the shortest deviation through j available, and is therefore possibly the second shortest path. The Hoffman and Pavely algorithm investigates all such paths (which satisfy the definition of the second shortest path) and therefore correctly identifies the second shortest path.

3.7. Murchland's All Paths Algorithm

For some applications all the possible paths between two specified nodes may be required. There are several ways of doing this and the one which follows is due to Murchland[7]. It should be remembered that in many networks the number of paths between a pair of nodes will be large and the corresponding computation required to find them will be long.

Murchland designates each arc by a letter and then describes an algebra for finding the paths. Consider the graph of Fig. 3.6a. There

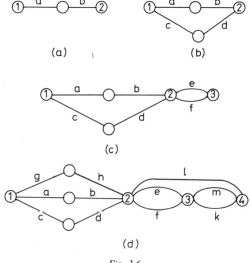

Fig. 3.6

is only one path between node 1 and node 2. It is designated 'ab' which is read 'a and b'. In Fig. 3.6b there is a second path cd and the existence of two paths between node 1 and node 2 is indicated by writing '$ab+cd$' which is read 'a and b or c and d'. In Fig. 3.6c note that there are two paths $(e+f)$ between node 2 and node 3, and four paths $(abe+abf+cde+cdf)$ between nodes 1 and 3. But note that $abe+cde+abf+cdf = (ab+cd)(e+f)$. This example shows that in Fig. 3.6c the paths between node 1 and node 3 can be found by multiplying the paths between nodes 1 and 2 by the paths between nodes 2 and 3. In Fig. 3.6d, more nodes and arcs have been added to show that all the paths between node 1 and node 4 are given by the product: $(gh+ab+cd)[(e+f)(m+k)+l]$.

For computational purposes we form an $N \times N$ matrix P in which the coefficient p_{ij} is equal to the sum (as defined above) of the known paths between i and j. Initially only the one-arc paths between nodes that are directly joined are assumed known. The matrix is then transformed step by step until at termination the element p_{ST} is equal to the sum of all paths between S and T. To avoid obtaining loops adopt the rule that the product of two path segments is zero if there is at least one arc common to both segments. For example $(cdl)(abl) = 0$, and $a(ab) = 0$. The actual steps of the algorithm now described will yield all paths from S to all other nodes, however it will be evident that a simple modification will yield all paths between all pairs of nodes.

Step 1 Entries in column 1 may be composed of more than one path. Each path is a *component* of the entry and may be *circled* or *uncircled*. Initially let all components be uncircled.

Step 2 Search for an uncircled component in column S (excluding entry p_{ST}). If one is found in some row i, call it q_{Si} and proceed to Step 3. Otherwise terminate.

Step 3 Search for an entry p_{ij} ($j \neq S$) which is not zero (an entry p_{ij} will be zero if no paths are known from node i to node j). For every such entry replace the value of p_{Sj} by the new value $p_{Sj} + q_{Si} p_{ij}$. When this operation has been carried out for all non-zero p_{ij}, circle component q_{Si} and return to Step 2.

When the algorithm terminates each entry p_{Sj} represents all the paths from S to j. In particular p_{ST} gives all paths between source and sink.

The following five tableaux (Fig. 3.8) show how the calculations proceed in finding all the paths between S and T in the network of

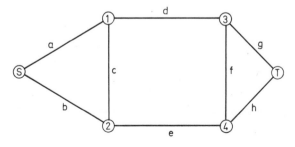

Fig. 3.7

58 Graphs and Networks

Fig. 3.7. (Note that changes occur only in the S column and so only it has been repeated.)
At the final iteration the last tableau shown is repeated but since all the entries are circled the computation terminates.

	S	1	2	3	4	T
S	0	a	b	0	0	0
1	a	0	c	d	0	0
2	b	c	0	0	e	0
3	0	d	0	0	f	g
4	0	0	0	f	0	h
T	0	0	0	g	h	0

0
ⓐ+bc
ⓑ+ac
ad
be
0

0
ⓐ+bc
ⓑ+ac
ⓐⓓ+bcd+bef
ⓑⓔ+ace+adf
adg+beh

0
ⓐ+bⓒ+befd
ⓑ+aⓒ
ⓐⓓ+bcd+beⓕ+acef
ⓑⓔ+ace+adⓕ+bcdf
adg+beh+bcdg+befg+aceh+adfh

0
ⓐ+bc+befⓓ+acefd
ⓑ+aⓒ+befdc
ⓐⓓ+bcd+bef+acef
ⓑⓔ+ace+adf
adg+beh+bcdg+befg+aceh+acefg+adfh

Fig. 3.8

EXERCISES

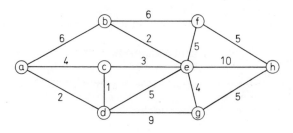

Fig. 3.9

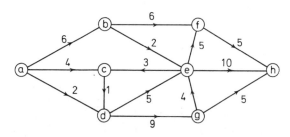

Fig. 3.10

1. For the two networks shown in Figs. 3.9 and 3.10 find the shortest path from a to h using the Dijkstra algorithm.
2. Repeat exercise 1 using the Ford algorithm.
3. For the network of Fig. 3.10 change the length of arc de to -5, and again attempt to find the shortest path from a to h using the Dijkstra and Ford algorithms. Repeat this exercise with a length of -4 for de.
4. For the networks of exercise 1, find the lengths of the shortest paths between all node pairs. (You can check any given row in your answer by using the Ford algorithm with the node corresponding to that row as source.)
5. For the networks of exercise 1, find the lengths of the second shortest paths from a to h using the Pollack algorithm.
6. For the networks of exercise 1, find the lengths of the second shortest paths from a to h using the algorithm of Hoffman and Pavley.
7. Using Murchland's algorithm, find all paths between a and h in the networks of Figs. 3.9 and 3.10.
8. Draw your own small networks, both directed and undirected,

having from eight to ten nodes, and repeat exercises 1, 2, 4, 5, 6, and 7 for them.

9. Prove that the Ford algorithm terminates in a finite number of steps providing there is no circuit in which the sum of arc lengths is negative.

10. Upon termination Floyd's algorithm will give a matrix showing the shortest distances between all node pairs. Invent your own notation which, when used in operating the algorithm, will identify the paths themselves.

11. Write a computer program for one of the algorithms of this chapter. If you are to solve large problems will the tableau methods of calculation described in the text be satisfactory?

12. The decanting problem is described in exercise 21 of Chapter 2. The graph used to solve the problem is shown in the answers. Suggest a means for finding a solution in the minimum number of pourings.

13. The travelling salesman problem is that of finding the shortest route for a salesman who must travel to each of n cities returning finally to the city from which he started. Can this problem be solved using one of the algorithms in this chapter? If not, explain why not.

14. In these days of rapid jet aircraft and crowded terminals actual flight time is becoming a smaller part of total travel time. In the airline network shown in Fig. 3.11, flight time is negligible in com-

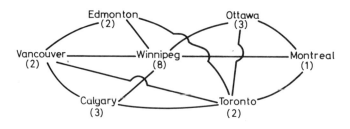

Fig. 3.11

parison to terminal time. The time (in hours) required to make a connection at each terminal is shown beside the city names. Suppose that you wish to travel from Vancouver to Montreal in the shortest possible time. Reformulate the network so that a shortest path algorithm can be used. Draw up a timetable showing the best (shortest) route between each city pair.

ANSWERS TO SELECTED EXERCISES

1. Fig. 3.9: *adcegh* (15 units), Fig. 3.10: *adgh* (16 units).
3. With a length of -5, the Ford algorithm will not operate as the total length of the circuit *cde* is -1. With a length of -4 the problem can be solved by the Ford algorithm. The Dijkstra algorithm does not operate if any arc lengths are negative.
5. Fig. 3.9: *acegh* (16 units); Fig. 3.10: *abfh* (17 units).
12. Draw all of the feasible arcs, and assign all arcs length 1. Join those nodes constituting a solution to a common sink by arcs of length zero. The shortest path through the network from initial node, called (0, 0), to the sink represents a solution in the smallest number of pourings.

REFERENCES

1. DIJKSTRA, D. W., 'A Note on Two Problems in Connexion with Graphs', *Numerische Mathematik*, **1** (1959).
2. DREYFUS, S. E., 'An Appraisal of Some Shortest Route Algorithms', *Operations Research*, **17**, no. 3 (May/June 1969).
3. FLOYD, R. W., 'Algorithm 97, Shortest Path', *Communications of the Association for Computing Machinery*, **5**, 345 (1962).
4. FORD, jr, L. R. and FULKERSON, D. R., *Flows in Networks*, Princeton (1962).
5. HOFFMAN, W. and PAVLEY, R., 'A Method for the Solution of the *N*th Best Path Problem', *J. Assoc. Computing Machinery* (1959).
6. MURCHLAND, J. D., 'A Bibliography of the Shortest Route Problem', *London Graduate School of Business Studies Report*, LBS-TNT-6 (Aug. 1967).
7. MURCHLAND, J. D., 'A New Method for Finding all Elementary Paths in a Complete Directed Graph', *London School of Economics Report*, LSE-TNT-22 (1965).
8. POLLACK, M., 'The *k*th Best Route Through a Network', *Operations Research*, **9** (1961).

CHAPTER 4

Flow Networks

4.1. Definitions

In this section we shall examine networks in which the nodes are connected by arcs through which some homogeneous quantity can flow. We refer to the flow in the arc joining node i to node j as $f(ij)$, and the set of flows $F = \{f(ij)\}$ in all arcs is called the flow function. We shall first treat those flows which do not vary with time and are then called steady state or static flows. Each arc of the network is limited in the amount of flow which it can carry. This limit $c(ij)$ is called the arc capacity. The set of all arc capacities $\{c(ij)\}$ is called the capacity function. The capacity function restricts the values that the flow function can take. In all arcs:

$$f(ij) \leq c(ij) \qquad (4.1)$$

4.2. The Conservation Equations

The networks treated contain a source node S from which all flow emanates, and a sink node T where all flow terminates. We assume that no flow enters the source or leaves the sink and that flow does not emanate or terminate at nodes other than S and T. Letting the amount of flow from S to T be v, these conditions can be written as conservation equations:

$$\sum_{j=1}^{N} f(Sj) = v \qquad (4.2)$$

$$\sum_{i=1}^{N} f(ix) - \sum_{j=1}^{N} f(xj) = 0 \qquad (4.3)$$

(for all nodes x, where $x \neq S$ and $x \neq T$)

$$\sum_{i=1}^{N} f(iT) = v \qquad (4.4)$$

In these summations we use $S = 1$ and $T = N$ for the sake of simplicity. Elsewhere we will continue to use S and T explicitly (as in $f(S_j)$ and $f(iT)$).

Any flow function which satisfies these four equations for a specific network is a feasible flow which means it can be realised in the network. We can now consider two problems. Firstly we can look for the value of the maximum feasible flow in a given network (the maximum permissible value of v), and secondly for a flow function having this value. We will now produce a theorem which solves the first problem and then deduce an algorithm which constructs a maximum flow function. Both the theorem and the algorithm are due to Ford and Fulkerson[1].

The Max-Flow/Min-Cut Theorem states that the value of the maximum feasible flow from S to T is equal to the capacity of the minimum cut (the sum of the arc capacities of the cut) separating S and T.

Consider a network having N nodes which are divided into two subsets X and $\bar{X}$ such that the arcs joining X and $\bar{X}$ form a cut separating S and T. If $S \in X$, then $T \in \bar{X}$. Let us now sum equation (4.2) for the flow at S and the set of equations (4.3) for the flow at the remaining nodes of X. The result is:

$$\sum_{x \in X} \left[\sum_{j=1}^{N} f(xj) - \sum_{i=1}^{N} f(ix) \right] = v$$

Since all nodes are either in X or $\bar{X}$ this can be expanded to the following form:

$$\sum_{x \in X} \left[\left\{ \sum_{j \in X} f(xj) + \sum_{j \in \bar{X}} f(xj) \right\} - \left\{ \sum_{i \in X} f(ix) + \sum_{i \in \bar{X}} f(ix) \right\} \right] = v$$

The first and third terms cancel to give:

$$\sum_{x \in X} \sum_{j \in \bar{X}} f(xj) - \sum_{i \in \bar{X}} \sum_{x \in X} f(ix) = v$$

This equation states that the net flow across the cut $(X\bar{X})$ is equal to v

From this it is evident that:
$$v \leq \sum_{x \in X} \sum_{j \in \bar{X}} f(xj) + \sum_{i \in \bar{X}} \sum_{x \in X} f(ix)$$
We also know that in order to satisfy (4.1) in each arc of the cut $(X\,\bar{X})$:
$$\sum_{x \in X} \sum_{j \in \bar{X}} f(xj) + \sum_{i \in \bar{X}} \sum_{x \in X} f(ix) \leq \sum_{i \in X} \sum_{j \in \bar{X}} c(ij)$$
therefore,
$$v \leq \sum_{i \in X} \sum_{j \in \bar{X}} c(ij) \tag{4.5}$$

This inequality must hold for every feasible flow and in each cut of the network. If this constraint is satisfied as an equality, clearly no larger flow and no smaller cut can exist.

Suppose now that v is a maximum flow. Define a set of nodes Z in the following way: let $S \in Z$, and in general if $x \in Z$ then $y \in Z$ if $f(xy) < c(xy)$ or if $f(yx) > 0$. All other nodes belong to $\bar{Z}$, the complement of Z. It will now be shown that the partitioning of the nodes of the network into $(Z\,\bar{Z})$ has defined a cut where if (ij) is an arc of the cut, $f(ij) = c(ij)$. Suppose for a moment that this is not so. Then $T \in Z$ and there is at least one S–T chain composed of arcs (xy) all of which satisfy either $f(xy) < c(xy)$ or $f(yx) > 0$. In this chain increase the flow in the arcs satisfying $f(xy) < c(xy)$ by a small amount δ and decrease the flow in the arcs satisfying $f(yx) > 0$ by the same amount. If δ is not too large (the algorithm that follows finds the largest admissible value) the resulting flow function satisfies constraints (4.1) through (4.4), and is therefore a feasible flow of value $v + \delta$. We did assume that v was a maximum flow. Therefore the flow of value $v + \delta$ cannot exist and the S–T chain described cannot exist. We can now conclude that $T \in \bar{Z}$ and $(Z\,\bar{Z})$ is a cut where $\sum_{\substack{i \in Z \\ j \in \bar{Z}}} f(ij) = \sum_{\substack{i \in Z \\ j \in \bar{Z}}} c(ij)$

then from (4.5) v is a maximum flow and $(Z\,\bar{Z})$ is a minimum cut.

We can now say with certainty that if there exists a flow if value v which is equal to the capacity of a cut $(X\bar{X})$ then the flow is maximum and the cut is minimum. Also if v is a maximum flow it is equal to the capacity of the minimum cut. This completes the demonstration of the theorem.

In proving this theorem we have seen that two types of flow augmenting chain can exist. In the first, all arcs (xy) of the chain have a flow $f(xy) < c(xy)$. In Fig. 4.1a, path S–b–a–T satisfies this criterion and the total flow can be increased by increasing the flow in each arc of the chain by one unit. In the second type of chain some arcs (xy) have a flow $f(yx) > 0$. In Fig. 4.1b S–b–a–T is such a chain. Total flow can be increased by raising the flow in S–b and

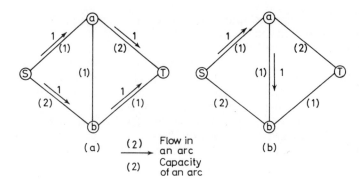

Fig. 4.1

a–T by one unit and diminishing the flow in b–a by one unit (preserving the conservation equations). These changes convert the one-chain flow of Fig. 4.1b to the two-chain flow of Fig. 4.1a (and the total flow can be increased again as above).

In the labelling algorithm which is able to discover both types of chain, nodes can be *labelled and unscanned, labelled and scanned*, or *unlabelled*. The meaning of these terms will become evident in the statement of the algorithm. There are two processes in the algorithm. The first searches for a flow augmenting path. If one is found, the second process shows how the flow function is modified to increase flow. If none is found the flow function at that point is optimal and the computation terminates.

4.3. The Ford-Fulkerson Maximum-flow Algorithm

The Labelling Process (a search for a flow augmenting path)
Step L1 Assign the label $(-, \infty)$ to the source which is now labelled and unscanned.
Step L2 Select any labelled and unscanned node x (initially the only such node is S). Suppose it to be labelled $(z^+, \mathcal{E}(x))$ or $(z^-, \mathcal{E}(x))$. If no such node can be found and the sink is unlabelled, terminate. Otherwise go to Step L3.
Step L3 Find an unlabelled node y directly connected to x. If no such node exists x is said to be scanned and return to Step L2. If such a node y is found and $f(xy) < c(xy)$, assign to y the label $(x^+, \mathcal{E}(y))$ where $\mathcal{E}(y) = \min\{\mathcal{E}(x), c(xy) - f(xy)\}$.
If on the other hand $f(yx) \geq 0$, assign to y the label $(x^-, \mathcal{E}(y))$, where $\mathcal{E}(y) = \min\{\mathcal{E}(x), f(yx)\}$.

This step is repeated until x is scanned. If at any point T becomes labelled go immediately to Step F1.

The Flow Change Process (construction of a new flow function)

Step F1 In general if node y is on the flow augmenting path and is labelled from x, then arc (xy) is an arc of the path. Since we know that T is the terminal node of the path, all arcs of the flow augmenting path may be found by tracing back from T.

Step F2 In those arcs (xy) of the flow augmenting path where y is labelled $(x^+, \mathcal{E}(y))$, increase the flow from $f(xy)$ to $f(xy) + \mathcal{E}(T)$. Where y is labelled $(x^-, \mathcal{E}(y))$, decrease the flow from $f(yx)$ to $f(yx) - \mathcal{E}(y)$.

Step F3 When all necessary flow changes have been made destroy the current labels and return to Step L1.

For small problems a solution tableau similar to that used for the Dijkstra algorithm can be constructed. Consider the network of Fig. 4.2. The initial tableau shown with the figure is the matrix of

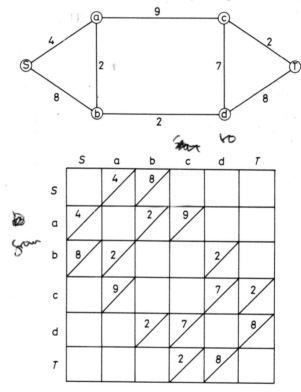

Fig. 4.2

arc capacities. In this case the matrix is symmetric because the network is not directed. Other networks may have assymetric matrices because they are directed or because arcs have a different capacity in one direction than in the other.

The squares of the tableau are divided into two parts, the upper

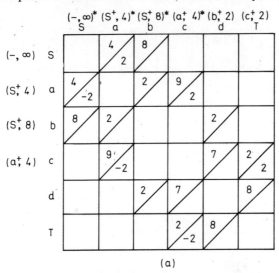

(a)

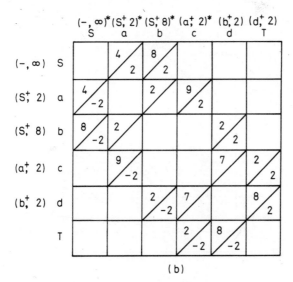

(b)

Fig. 4.3

68

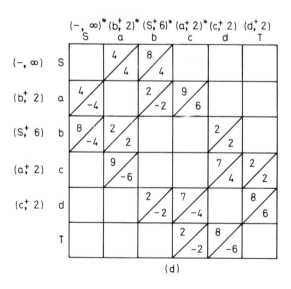

(c)

(d)

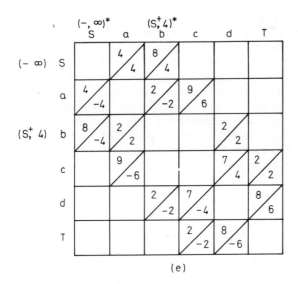

(e)

portion giving the arc capacity and the lower portion giving the flow. A flow of f units from y to x will be recorded as $+f$ in the yx square of the matrix and as $-f$ in the xy square. This permits easy recognition in Step L3 of the algorithm of the case where $f(yx) > 0$, requiring a label of the form $(x^-, \mathcal{E}(y))$ to be affixed to y. In Step F2 if the flow entry in square xy is increased by a certain amount, the same amount is subtracted from the already negative entry in square yx and inversely.

Labels are written across the top of the matrix, above the node names. When a node is chosen as the node x of Step L2 the label is also noted on the left hand side of the matrix (this helps labelling). When a node is scanned an asterisk is affixed to its label.

The problem of Fig. 4.2 is solved in five applications of the labelling process. The tableaux are shown in Fig. 4.3.

All calculations in this example have been carried out in integers. An examination of the algorithm shows that if the arc capacities are integers all flows are integers because $\mathcal{E}(y)$ will always have an integer value (Step L3).

It will be remembered from the proof of the max-flow/min-cut theorem that if maximum flow exists in a network there exists a cut $(Z\bar{Z})$ in which all arcs are saturated (that is to say $f(ij) = c(ij)$). At the end of the final iteration of the algorithm some nodes will be labelled (including S) and some unlabelled (including T). We can show that the labelled nodes correspond to the set Z and the unlabelled nodes to the set $\bar{Z}$. All the arcs joining labelled to unlabelled nodes must be saturated or both nodes would be labelled. We can also see that such arcs form a cut or there would exist at least one new flow augmenting path from S to T. By the max-flow/min-cut theorem this cut is therefore minimum and we are assured that the minimum cut lies between the set of labelled and the set of unlabelled nodes at the termination of the algorithm.

If the network is planar a dual network exists and another method can be used to locate the minimum cut. Let the length of an arc in the dual network be numerically equal to the capacity of the corresponding arc in the primal. From the previous chapter we know that each cut in the primal corresponds to a path in the dual. The minimum cut in the primal network therefore corresponds to the shortest path through the dual.

If there is more than one cut having the same minimum capacity all will be saturated. The maximum flow algorithm will locate the cut nearest to the source, while which cut is located through the use of the dual will depend on how the labelling is done in the shortest path algorithm.

4.4. Capacitated Nodes

In many real problems nodes as well as arcs have capacity limits. A road network is obviously such a case as both road segments and intersections have limited capacity. Maximum flow problems in networks of this type can be solved using the Ford-Fulkerson algorithm if the network is modified as follows. In directed networks each node i is divided into two nodes i_1 and i_2. These nodes are connected by an arc which is directed from i_1 to i_2 and has the capacity of the original node it replaces. All arcs which can carry flow into i in the original network are incident to i_1 after modification and arcs which can carry flow away from i are incident to i_2. A small directed network is shown in Fig. 4.4 both before and after

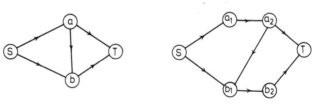

Fig. 4.4

modification. In undirected networks an additional step is necessary. Each undirected arc is replaced by two directed arcs orientated in opposite directions (arcs incident to S and T need not be doubled as flow neither enters the source nor leaves the sink) and the above procedure is carried out. An illustration is given in Fig. 4.5.

While these modifications permit us to solve problems in networks with capacitated nodes the size of the network is greatly

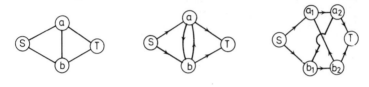

Fig. 4.5

increased, particularly if the original network is undirected. R. D. Wollmer[8] has published an algorithm to solve maximum flow problems in networks having both node and arc capacities. While more complex than the original Ford-Fulkerson algorithm it requires less computer storage space than the network transformation methods mentioned before.

4.5. Multiple Sources and Sinks

Some networks have more than one source and more than one sink. The flow algorithms described in this chapter can be directly applied to these networks if the following modifications are made. Add to the network an artificial master source and join it to all the real sources by arcs which are unrestricted in capacity. Add also an artificial master sink and join it to the real sinks in the same way

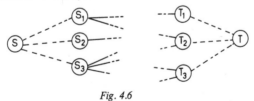

Fig. 4.6

(Fig. 4.6). If the maximum flow algorithm is then applied to this network, at termination the flows in the new unrestricted arcs show the amounts of flow sent out by each real source and received by each real sink.

4.6. Maximal Dynamic Flow

In some networks not only the maximum arc capacities $c(ij)$ but also the arc traversal times $t(ij)$ must be considered. An interesting problem is that of finding the maximum flow which the network can carry from source to sink in some given length of time P. In addition, flow received at a node can either be moved immediately or held over to a later time period.

This problem could be solved using the Ford-Fulkerson algorithm if the network were expanded as follows. Consider the network of Fig. 4.7. The figures on the arc represent the traversal times, the

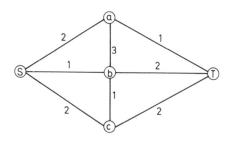

Fig. 4.7

capacities being unimportant for the moment. Suppose we wish to find the maximum flow for $P = 4$. Each node of the original network appears in the expanded network (Fig. 4.8) at the beginning of each time period. For example S expands to S_0, S_1, S_2, S_3 and S_4. Arcs in the expanded network are defined as follows: nodes x_{p_1} and y_{p_2} in the expanded network are joined if (xy) is an arc of the original network, if $p_2 - p_1 = t(xy)$ and if the arc $(x_{p_1} y_{p_2})$ can be part of some chain joining the source to the sink. In Fig. 4.8 we note that S_0 and a_2 satisfy these criteria and are joined. However S_0 and a_1 are not joined because $t(Sa) = 2$ and S_2 is not joined to a_4 because this arc cannot be part of any chain reaching a sink node at or before $P = 4$. Arcs such as $(S_0 S_1)$ which permit flow to be held over at a node are assigned an infinite capacity. All arcs are directed regardless of whether or not the original network was directed because time is irreversible.

Because of the way in which the network of Fig. 4.8 was constructed, solving the static maximal flow problem (using the technique for multiple sources and sinks) in this network is equivalent to solving the dynamic maximal flow problem in the original network. To expand a network in this way for the solution of a dynamic

Flow Networks   73

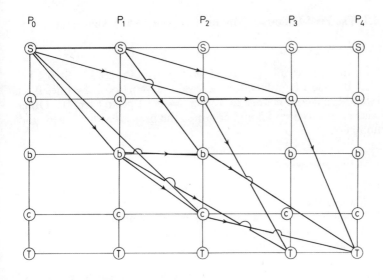

Fig. 4.8

flow problem is cumbersome and impractical if the original network is large or the number of time periods great. However since solution in this manner is possible we gather that the problem of finding a maximal dynamic flow in a network is in principle the same as that of finding a maximal static flow. What we require to solve the problem are new methods of computation. Ford and Fulkerson[2] have provided us with an algorithm which constructs maximal dynamic flows from static flows.

The algorithm has a labelling routine and a flow change routine both which resemble those of the static flow algorithm. However a control variable π_i (which is an integer associated with the node i) is used to ensure that the constraints imposed by the arc traversal times are respected. A time advance routine is included to adjust the values of the π_i. These three routines produce a static flow which is then decomposed into separate chain flows by a fourth routine. To obtain the maximal dynamic flow over the period P it has been shown that we need only repeat the chain flows found by this method as often as possible within the time limit.

In the algorithm arcs (ij) for which $\pi_i + t(ij) = \pi_j$ will be called admissible, and only if (ij) is admissible can j be labelled from i. Nodes are unlabelled, labelled and unscanned, or labelled and scanned as in the static algorithm.

4.7. The Ford-Fulkerson Maximal Dynamic Flow Algorithm

Control Steps

C1 Initially let all $\pi_i = 0$ and all $f(ij) = 0$. Proceed to Step L1.
C2 If node i is unlabelled increase the value of π_i by one. If i is labelled make no change in π_i. If $\pi_T = P+1$ go to Step D1. Otherwise return to Step L2 and continue the process with the new node numbers.

Labelling Process

L1 Attach the label $(-^+, \infty)$ to the source which is now labelled and unscanned.
L2 Select any labelled unscanned node x (initially only S satisfies this criterion) and suppose it to be labelled $(z^+, \mathcal{E}(x))$ or $(z^-, \mathcal{E}(x))$. If no such node x can be found and the sink is unlabelled go to Step C2.
L3 Find an unlabelled node y directly connected to x and such that the arc (xy) is admissible. If such a node y is found and $f(xy) < c(xy)$ assign to y the label $(x^+, \mathcal{E}(y))$ where $\mathcal{E}(y) = \min\{\mathcal{E}(x), c(xy) - f(xy)\}$. Also, to nodes y such that (yx) is admissible and $f(yx) > 0$ assign the label $(x^-, \mathcal{E}(y))$ where $\mathcal{E}(y) = \min\{\mathcal{E}(x), f(yx)\}$. When all nodes directly connected to x have been examined, x becomes scanned; return to Step L2. However if at any point T becomes labelled go immediately to the flow change process.

Flow Change Process

The steps of the flow change process are identical to those listed for the static algorithm. Upon completion of this process go to Step L1.

Decomposition Process

D1 Assign to S the label $(-, \infty)$ and considers S to be unscanned.
D2 Take any labelled unscanned node x (initially only S satisfies this criterion). Suppose that it is labelled $(z, \mathcal{E}(x))$. To all nodes y that are unlabelled and such that $f(xy) > 0$, assign the label $(x, \mathcal{E}(y))$ where $\mathcal{E}(y) = \min\{\mathcal{E}(x), f(xy)\}$. When x becomes scanned, find another labelled unscanned node and repeat D2. If T becomes labelled at any point go to D3.
If T is unlabelled and no new labels can be assigned, terminate.
D3 Suppose T is labelled $(k, \mathcal{E}(T))$. Trace the chain back from sink to source (as in the flow change process) using the node labels. Reduce the flow in each arc of the chain by $\mathcal{E}(T)$. Note the arcs

of the chain and the amount of flow that it carries ($\mathcal{E}(T)$). Remove all old labels and return to Step D1.

A short example will illustrate how the algorithm functions. Fig. 4.9 shows a network having five nodes and eight arcs. The

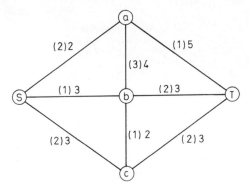

Fig. 4.9

figures in brackets are the arc traversal times while the remaining figures give the arc capacities. The maximum flow for $P = 5$ will be found.

The four applications of the labelling process and Step C2 shown immediately below are necessary before any change can be made.

Iteration	1		2		3		4	
node	π	label	π	label	π	label	π	label
S	0	$(-, \infty)$	0	$(-, \infty)$	0	$(-, \infty)$	0	$(-, \infty)$
a	0		1		2		3	
b	0		1	$(S^+, 3)$	1	$(S^+, 3)$	1	$(S^+, 3)$
c	0		1		2	$(S^+, 3)$	2	$(S^+, 3)$
T	0		1		2		3	$(b^+, 3)$

At this point a chain of admissible arcs (S–b–T) becomes available to carry flow from source to sink. The flow function becomes:

Arc	Sa	Sb	Sc	aT	ba	bT	cb	cT
Flow	0	3	00	0	0	3	0	0

After making the flow change the calculation recommences at Step L1. Note that no change is made in the node numbers at this point. Another attempt in iteration 5 is made to find a new flow augmenting path.

Node	Iteration 5	
	π	label
S	0	$(-, \infty)$
a	2	$(S^+, 2)$
b	1	
c	2	$(S^+, 3)$
T	3	$(a^+, 2)$

The attempt is successful and the flow function is changed to:

Arc	Sa	Sb	Sc	aT	ba	bT	cb	cT
Flow	2	3	0	2	0	3	0	0

Iteration 6 using the same node numbers, does not yield another flow augmenting path and so the π_i are changed and iteration 7 is carried out.

Iteration	6		7	
Node	π	label	π	label
S	0	$(-, \infty)$	0	$(-, \infty)$
a	2		3	
b	1		2	
c	2	$(S^+, 3)$	2	$(S^+, 3)$
T	3		4	$(S^+, 2)$

Following iteration 7 the flow function is altered to:

Arc	Sa	Sb	Sc	aT	ba	bT	cb	cT
Flow	2	3	2	2	1	3	0	2

Flow Networks 77

In iteration 8 we re-label, finding no new path. The node numbers are changed once more in iteration 9, causing the path S–c–b–a–T to become admissible.

Iteration	8		9	
Node	π	label	π	label
S	0	$(-, \infty)$	0	$(-, \infty)$
a	3		4	$(b^+, 1)$
b	2		3	$(c^+, 1)$
c	2	$(S^+, 1)$	2	$(S^+, 1)$
T	4		5	$(a^+, 1)$

A flow change is again made.

Arc	Sa	Sb	Sc	aT	ba	bT	cb	cT
Flow	2	3	3	3	1	3	1	2

In iteration 10 an attempt is made to find another path with the existing node numbers but none is found and the algorithm proceeds to Step C2 where the numbers are again increased. We find that after the increase $\pi_T = 6$, indicating to us that the search for new flow augmenting paths must stop. The algorithm now goes to Step D1.

Iteration	10		11	
Node	π	label	π	label
S	0	$(-,)$	0	$(-,)$
a	4		5	
b	3		4	
c	2		3	
T	5		6	

The final flow function is now decomposed into chain flows each of which is repeated as often as possible in the time available. The table below summarises the results.

Chain	Chain flow value	Total traversal time	Times repeated	Total flow in chain
S–b–T	3	3	3	9
S–a–T	2	3	3	6
S–c–T	2	4	2	4
S–c–b–a–T	1	5	1	1

78   Graphs and Networks

The total flow which can be received by the sink in five time periods is 20.

It has been shown[3] that from the calculations giving the maximal dynamic flow for P periods the maximal dynamic flow for any lesser span of time can be found. It is only necessary to repeat the chain flows found for P periods as often as possible in the shorter time. In the problem solved above, for example, the maximal dynamic flow for four periods is obtained by using the chain flows shown in this table:

Chain	Chain flow value	Total traversal time	Times repeated	Total flow in chain
S–b–T	3	3	2	6
S–a–T	2	3	2	4
S–c–T	2	4	2	2

These chains are illustrated in Fig. 4.8.

4.8. Minimum Cost Maximum Flow

From previous problems it will be noted that it is common for maximum flow problems to have more than one optimum solution. This means that more than one feasible flow function yields the optimal value. In such cases (if there is a shipping cost as there often is) we would naturally wish to find the maximum flow function which minimises this. Ford and Fulkerson have described an algorithm[3] which will do this in the case where the shipping costs are linear.

Let the cost of shipping one unit of flow from node x to node y be $a(xy)$. The problem can then be stated as follows:

$$\text{minimise} \left\{ \sum_x \sum_y a(xy) f(xy) \right\}$$

subject to:

$$f(xy) \leq c(xy) \quad \text{(for all } x \text{ and } y\text{)}$$

$$\sum_{j=1}^{N} f(Sj) = \sum_{i=1}^{N} f(iT) = v$$

(The maximum flow is v and is equal to the flow out of the source and into the sink.)

$$\sum_{i=1}^{N} f(ix) - \sum_{j=1}^{N} f(xj) = 0$$

Flow Networks 79

(This holds for all x. These are the conservation equations.)
Define now a set of variables $\{\pi(i)\}$ called the node numbers. The initial values for these variables may be chosen arbitrarily and are adjusted by the algorithm. Using the node numbers define a set of arc numbers $\bar{a}(ij) = a(ij) + \pi(i) - \pi(j)$. Each arc (ij) is then in one of the following states:

(1) $\bar{a}(ij) > 0$ $f(ij) = 0$
(2) $\bar{a}(ij) = 0$ $0 \leq f(ij) \leq c(ij)$
(3) $\bar{a}(ij) < 0$ $f(ij) = c(ij)$
(4) $\bar{a}(ij) < 0$ $f(ij) < c(ij)$
(5) $\bar{a}(ij) > 0$ $f(ij) > 0$

Each arc is said to be 'in kilter' or 'out of kilter' which means it is in a state in which one could expect to find it in a minimal cost solution or it is not. We will now define 'kilter numbers', which are measures of how far an arc is from being in 'kilter'.

States (1), (2) and (3) are 'in kilter' states and the corresponding kilter number is zero. For state (4) the kilter number is

$$\bar{a}(ij)\{f(ij) - c(ij)\},$$

and for state (5) the kilter number is

$$\bar{a}(ij)\{f(ij)\}.$$

It is easy to see that the kilter numbers are always non-negative.

We will proceed by first finding a maximum feasible flow and if it is not already of minimum cost, altering it so that the flow function is still both feasible, maximal and of lower cost than the original. This can be done by selecting an out-of-kilter arc and changing the flow in it to lower the kilter number. The flow in other arcs must be altered in order to retain feasibility and maximality. Consideration of the conservation equations shows that the set of arcs altered must form a cycle. After an out of kilter arc has been selected the algorithm will label through the network in search of a cycle in which flow may be altered. Having found one the alterations are made in such a way that the kilter number of the selected arc is reduced and no other kilter number is increased.

4.9. The 'Out of Kilter' Algorithm for Minimal Cost Maximum Flow

The algorithm starts with a maximum flow already existing.
Step 1 Locate an out-of-kilter arc pq. If none can be found, terminate, as the current solution is optimal. If the arc found is in state (4) go to Step 2. If it is in state 5 go to Step 3.
Step 2 Assign to q the label $\{p^+, \varepsilon(q) = c(pq) - f(pq)\}$. Go to Step 4.

Step 3 Assign to p the label $\{q^-, \mathcal{E}(p) = f(pq)\}$ and go to Step 4.
Step 4 Label the remaining nodes of the network according to the following two rules:
  (a) If node x is labelled $\{z^{\pm}, \mathcal{E}(x)\}$, y is unlabelled, and $\bar{a}(xy) \leq 0$ where $f(xy) < c(xy)$, then node y may be labelled with $\{x^+, \mathcal{E}(y) = \min(c(xy) - f(xy), \mathcal{E}(x))\}$.
  (b) If node x is labelled $\{z^{\pm}, \mathcal{E}(x)\}$, y is unlabelled, and $\bar{a}(yx) \geq 0$, where $f(yz) \geq 0$, then node y may be labelled with $\{x^-, \mathcal{E}(y) = \min(\mathcal{E}(y), f(yx) - c(yx))\}$.

Carry out these steps until both p and q are labelled. At this point proceed to Step 5. If no further labels can be placed and either p or q is unlabelled (depending on whether the arc being treated is in state (4) or state (5)) go to Step 6.

Step 5 (Flow change Step). The flow is altered in the circuit found in Step 4.
  (a) If the arc pq was in state (4), $\mathcal{E}(p)$ is added to the flow in the forward arcs of the path and subtracted from the flow in the reverse arcs. The quantity $\mathcal{E}(p)$ is added to the flow in pq.
  (b) If pq was in state (5), $\mathcal{E}(q)$ is subtracted from $f(pq)$ and the flows in the remaining arcs of the circuit are altered by adding or subtracting $\mathcal{E}(p)$ from existing flows as above. The new kilter numbers are now calculated. If pq is not yet in kilter go to Step 2 and repeat the procedure. Otherwise go to Step 1, choose another out-of-kilter arc and continue.

Step 6 (Change of node numbers). Let X be the set of labelled nodes and Y the set of unlabelled nodes (X and Y are mutually exclusive and exhaustive sets). Define two subsets of arcs:

$$A_1 = \{(xy)/x \in X, y \in Y, \bar{a}(xy) > 0, f(xy) \leq c(xy)\}$$
$$A_2 = \{(yx)/x \in X, y \in Y, \bar{a}(yx) < 0, f(yx) \geq 0\}$$

Let $\delta_1 = \min_{A_1} (\bar{a}(xy))$, $\delta_2 = \min_{A_2} (-a(yx))$, and $\delta = \min(\delta_1, \delta_2)$.

(If A_i is empty, let $\delta_i = \infty$.)
For all $y \in Y$, let the new node number for y be $\pi(y) + \delta$.
Return to Step 1.

4.10. Increasing the Capacity of a Network

Previously we have examined the problem of increasing the capacity of a network having capacitated arcs. We can also consider the problem which arises when this maximum flow is attained and judged to be insufficient. We must now increase the capacity of the network. It is normal to assume that such changes cannot be made without cost and that in general the cost per unit of capacity increase is different for each arc.

If the cost per unit increase in the capacity of an arc is constant over the range of the increase (a linear cost function) the problem can be solved using a modified version of the preceding algorithm

In reality the cost to increase the capacity of a given arc is more likely to be non-linear. In many cases it may be found to have the form of a step function (Fig. 4.10). For a given expenditure a certain

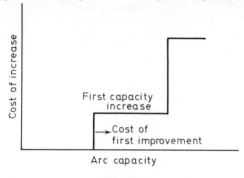

Fig. 4.10

capacity increase can be obtained, but for less than that nothing can be accomplished. An example of this can be found in railway operations where if a given stretch of track is operating at maximum capacity, capacity may be increased by building a second track. This may double capacity but it is not possible to obtain half the increase by an expenditure of half the amount[5,6].

EXERCISES

1. Find the maximum flow from S to T in the directed network of Fig. 4.11.

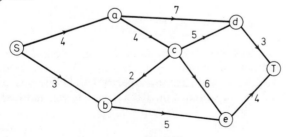

Fig. 4.11

2. In the network of exercise 1, find the maximum flow from S to T this time starting the calculation with an existing flow of 4 units in the chain S–a–c–e–T.

3. Find the maximum flow from S to T in the undirected network of Fig. 4.12. Identify the minimum cut and verify that after the final iteration its arcs lie between labelled and unlabelled nodes.

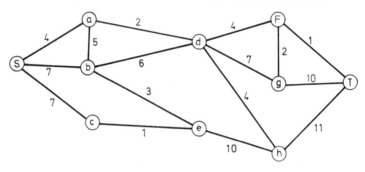

Fig. 4.12

4. Find the dual of the network of Fig. 4.12 and use it to locate all cuts having the minimum value. (You can use Pollack's algorithm given in Chapter 3).

5. Find the maximum flow from S to T in the network of Fig. 4.13. Note that some of the nodes are capacitated while others are not.

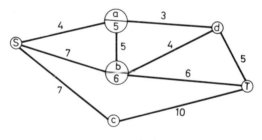

Fig. 4.13

6 (a). In Fig. 4.11, let node c have the capacity 5 and node d have the capacity 2. Find the maximum flow from S to T in this network.

(b). In Fig. 4.12, let node d have the capacity 7 and node e have the capacity 5. Find the maximum flow from S to T in this network.

7. Find the maximum flow that can reach T in the network of Fig. 4.14.

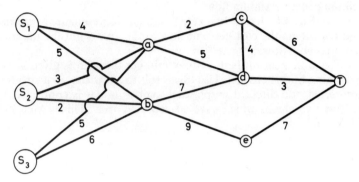

Fig. 4.14

8 (a). In the network of Fig. 4.15, let the numbers in brackets represent the arc traversal times and the others the arc capacities. Find the maximal dynamic flow for 4, 5 and 6 time periods.

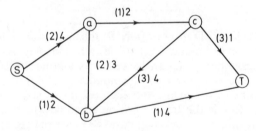

Fig. 4.15

(b). In Fig. 4.16, let the bracketed numbers represent the capacities and the others the arc traversal times. Again find the maximal dynamic flow for 4, 5 and 6 periods.

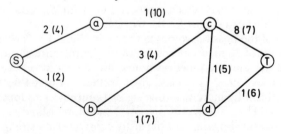

Fig. 4.16

9 (a). In Fig. 4.15, let the bracketed numbers represent the cost per unit flow in the arcs, and the others the arc capacities. Find the minimal cost maximum flow.

(b). In Fig. 4.16, let the bracketed numbers represent arc capacities and the others the cost per unit flow in the arcs. Find the minimal cost maximal flow.

10. The statement of the max-flow/min-cut theorem is given at the beginning of this chapter. Use this result to show that in a strongly connected non-directed graph, each pair of nodes is connected by at least two chains which have no common arcs. (This is Menger's Theorem.)

11.

		1 C_1	6 C_2	2 C_3	6 C_4
5	W_1	5	4	3	2
5	W_2	10	8	4	7
5	W_3	9	9	8	4

The Hitchcock or Transportation Problem is well known and efficient solution techniques for it are described in all texts on mathematical programming[7]. It can be stated as follows: given a set of warehouses w containing known amounts of goods, customers c having known requirements, and the cost a_{ij} to ship one unit of goods from i to j, find a transportation schedule which satisfies all customers at minimum shipping cost. The cost matrix for a small example is shown above (the availabilities and requirements are shown in the margins of the matrix). Formulate this as a network problem and solve it using the methods of this chapter.

12. An army commander is required to deploy a large number of troops in a war zone as quickly as possible. His base is served by road, rail, and canal which connect him with air and sea ports from which he can depart for the battle field.

A map of the possible routes is shown in Fig. 4.17. The capacities (in platoons of 36 men) of each leg of the journey are shown on the map and the journey times (in hours) are shown in brackets. Note that because of a regulation laid down by the Air Force Pilots Union, no aircraft can fly more than one leg of a journey. This forces transshipment of some soldiers at Airport 2 and at the island base. Draw up a histogram showing how the army commander's strength in the field will build up over the first 48 hours of the operation.

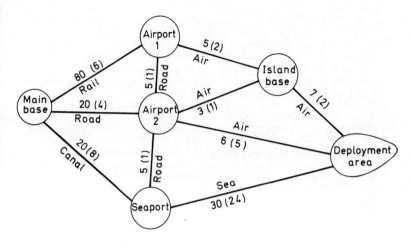

Fig. 4.17

ANSWERS TO SELECTED EXERCISES

1. The maximum flow is 7 units.
2. The maximum flow is 12 units. The Ford-Fulkerson algorithm will identify arcs Sa, Sb and ce as being the minimum cut.
4. This exercise will identify the cut formed by arcs ad, bd, be and ce as being a second cut of minimum value.
5. The maximum flow is 16.
6 (a). The maximum flow is 6.
 (b) The maximum flow is 11.
8 (a).

number of time periods	chains used	times repeated	total flow
4	S–b–T (2 units)	3	6
5	S–b–T (2 units) S–a–b–T (2 units)	4 1	10
6	S–b–T (2 units) S–a–b–T (2 units) S–a–c–T (1 unit)	5 2 1	15

9 (b). Chain $S-a-c-b-d-T$ (4 units)
 Chain $S-b-d-T$ (2 units)
 Total cost of 38 units.

REFERENCES

1. FORD jr, L. R. and FULKERSON, D. R., *Flows in Networks*, Princeton (1962).
2. FORD jr, L. R., and FULKERSON, D. R., 'Constructing Maximal Dynamic Flows from Static Flows', *Operations Research*, **6**, no. 3 (May/June 1958).
3. FORD jr, L. R. and FULKERSON, D. R., 'An Out-of-Kilter Algorithm for Minimal Cost Flow Problems', *J. Soc. Indust. Appl. Math.*, **9** (1961).
4. FULKERSON, D. R., 'Increasing the Capacity of a Network: The Parametric Budget Problem', *Management Science*, **5**, no. 4 (1959).
5. HAMMER, P. L., 'Increasing the Capacity of a Network', *Can. O. R. Journal*, **7**, no. 2 (1969).
6. PRICE, W. L., 'Increasing the Capacity of a Network Where the Costs are Non-Linear: A Branch-and-Bound Algorithm', *Can. O.R. Journal*, **5**, no. 2 (1967).
7. VAJDA, S., *Mathematical Programming*, Addison-Wesley (1961).
8. WOLLMER, R. D., 'Maximising Flow Through a Network with Node and Arc Capacities', *Transportation Science*, **2**, no. 3 (1968).

CHAPTER 5

Activity Networks

5.1. Introduction

The term *activity network* is used to describe various project planning techniques such as PERT[1,6], CPM[1,6,4] and Roy's Method of Potentials[1,8,9] (also called Precedence Diagrams). These techniques each have separate features but all are methods of planning through network models.

In PERT and CPM arcs of a network are used to represent activities (jobs, collections of jobs or portions of jobs) and the nodes represent events (start or finish or the completion of some important job phase). The precedence relationships among jobs are shown by the structure of the network. In the method of potentials nodes are used to represent jobs and arcs indicate the sequential relationships between jobs. The length assigned to an arc is equal to the time which must pass between the start of successive jobs. It is claimed that this method of drawing the network allows any changes in a project to be more easily incorporated in the model.

As an example let us make a network model of an Operational Research study of a system. The following activities are judged to be necessary:
  (A) Make a preliminary study of the structure of the system
  (B) Construct a theoretical model
  (C) Collect data
  (D) Encode the data for the model constructed
  (E) Test the model with artificially generated data

(F) Present the model to scientific colleagues for evaluation
(G) Present the model to the user for evaluation
(H) Revise as necessary
(I) Test the model with real data
(J) Final revisions
(K) Present the model for use

Note that some of these activities can go on concurrently and others can not. The project can be described by the network shown in Fig. 5.1. The arcs shown by broken lines do not represent activities but show precedence relationships only. The use of these 'dummy' arcs will be explained in a later section.

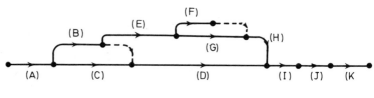

Fig. 5.1

The network in Fig. 5.1 is of the PERT and CPM type. The Roy network for the same project is shown in Fig. 5.2. Note that two artificial jobs, to represent lead time and termination, have been added to the network. This is often necessary in order to obtain a network having unique initial and terminal nodes.

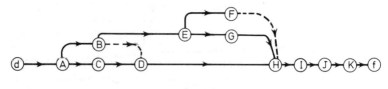

Fig. 5.2

Such network models are quite valuable in the planning stages of complex projects where they can serve in the allocation of resources and in the setting up of work schedules. They can also be used during the execution of the project to aid the control of various jobs, and for steering the project round delays and avoidable expenditures.

The remainder of this chapter will deal with the use of PERT and CPM models. For a description of Roy's method consult the references already given.

5.2. Networking Rules and Definitions

In the previous section an example of an activity network was given. At this point it is necessary to formalise the rules for drawing such networks from the list of activities making up the project.

In a network an activity is any procedure which takes time. It may be necessary to break up a given activity into several parts because of the individual importance of each part. In other cases several activities can be represented by one arrow, because although they are separate operations they must be done in known sequence and without interruption (for example the operations of heating and forming a metal part). Activities are represented in the network by arrows which are assigned lengths equal to the completion time required but note that the arcs are not drawn to scale.

An event is the beginning or end of an activity and is represented by a node in the network. One event can represent at the same time the end of several activities and the beginning of several others. For example in Fig. 5.1 a single event represents the end of activities (D) and (H) and the beginning of activity (I).

The arrows are drawn in the network in such a way as to show the logical sequence in which jobs must be done. In Fig. 5.1 it is obvious that activity (D) (encode data) must follow activity (C) (collect data). Note that it may eventually happen that two jobs must be done in sequence rather than simultaneously for some reason unrelated to the jobs themselves (when each job is done by the same individual), however it is important that only those precedence relationships which exist because of the nature of the jobs themselves be included in the initial network drawn.

Dummy activity arcs are used to show activity sequences in certain situations. In the previous example before commencing to Encode Data it is necessary not only to Collect Data but also to Construct Model because this will determine the form in which the data must be encoded. This situation could be shown as in Fig. 5.3a but in activity networks a given pair of nodes must identify one activity and one only. Therefore a dummy arc having an execution time of zero is inserted into the network as shown in Fig. 5.3b to permit the logical sequence to be demonstrated and the network rules to be respected.

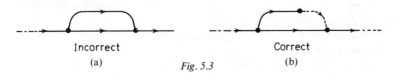

Incorrect
(a)
Fig. 5.3
Correct
(b)

Networks must have a unique initial and terminal node. The network shown in Fig. 5.4a is incorrectly drawn. The situation is corrected in Fig. 5.4b by the addition of a lead-in activity and an ending activity. Often these activities may be real as in Fig. 5.1, but if required they may be dummies.

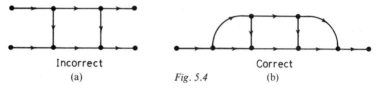

Incorrect (a)   *Fig. 5.4*   Correct (b)

Activity networks are completely directed because time is not reversible, and for the same reason must contain no directed cycles. The existence of the directed cycle shown in Fig. 5.5 would imply

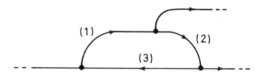

Fig. 5.5. An inadmissible cycle

that it is possible to complete activity (1) activity (2) and activity (3) (each of which has a non-zero completion time) in such a way that the beginning of activity (1) coincides with the end of activity (3). This implies that it is possible to complete three real activities in zero time. It may happen that a directed cycle is mistakenly included in the initial version of a complex network. Such a loop must be eliminated (computational methods exist for this) because it is clearly an error in the model.

5.3. Fulkerson's Node Numbering Scheme

Up to now we have referred to activities by number or by description. It is common however to refer to activities by their initial and terminal nodes. Fulkerson[2] has described a scheme for numbering events in such a way that for any activity beginning with event i and ending with event j, $i < j$.

Step 1 Number the origin, which has only outward pointing arcs, as one. Let $i = 1$.

Step 2 Delete all arcs leading out of newly numbered nodes.

Step 3 Search for nodes in this modified network which have only outward pointing arcs. Number these nodes (in any order) from $i+1$ to $i+k$. Let the new value of i be $i+k$ and return to Step 2. When all nodes have been numbered the computation is terminated.

Some practitioners advise against trying to number events in this way. If a network is modified or revised so that some activities are deleted and new ones included the events must be completely renumbered. This disadvantage can be partially overcome by initially numbering events with a spaced sequence (for example 1, 5, 10, . . .). This leaves a degree of flexibility so that all events may not have to be re-numbered if slight changes are made.

One advantage of using the sequential method of numbering (even if no subsequent use is made of it) is that it provides a way of detecting directed cycles. If a directed cycle exists the event numbering procedure cannot number all nodes and will terminate prematurely. Consider the network of Fig. 5.6. After the initial event has been numbered (one) and the arcs leading from it deleted, no further events can be numbered as there exists no node which does not have arcs directed into it.

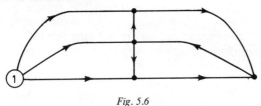

Fig. 5.6

On balance the sequential node numbering scheme is initially useful and some attempt to arrange event numbers so that the lowest are to the left and the highest to the right is useful in visualising the rough position of an event. However no computational scheme *requiring* events to be numbered in this way should be constructed unless the user is willing to tolerate the difficulties which will be encountered when making changes to the original network.

In preparing to draw an activity network from the job list of a project it may be useful to first draw a skeleton network. This skeleton should contain only the principal activities, and from it a feel for the form of the project can be gained.

5.4. The Critical Path

Since we wish to use activity networks to aid planning and control one of the factors of interest is the shortest time in which a project can be completed. This time is given by the longest path through the network (more than one path can have this same maximum length). This path is called the critical path and the arcs of the path represent critical activities.

The critical path or paths can be found in several ways. One is to use the algorithm shown below. This algorithm and the others similar to it depend on the fact that the networks have no directed cycles. In the algorithm each node j is assigned a two part label $(i, p(j))$ as in the shortest path algorithms of the previous chapter. The time taken to complete activity (ij) is $a(ij)$.

5.5. Algorithm for Finding the Critical Path

Step 1 Initially let all nodes bear the label $(-,0)$.
Step 2 Search for a directed arc (ij) such that $p(i)+a(ij) > p(j)$. If no such arc exists terminate. Otherwise proceed to Step 3.
Step 3 Replace the label on node j with the label $\{i, (p(i)+a(ij))\}$. Return to Step 2.

When the algorithm has terminated the critical path can be traced back through the labels (as in the shortest path algorithms of the previous chapter) and all critical jobs located.

As an example let us consider the project illustrated in Fig. 5.1. The activity times are as follows:

Activity	Completion time (weeks)	Initial and terminal event numbers
(A) Preliminary study	3	1,2
(B) Construct model	5	2,3
(C) Collect data	6	2,7
(D) Encode data	1	7,8
(E) Test with artificial data	2	3,4
(F) Criticism obtained from colleagues	2	4,5
(G) Criticism obtained from user	1	4,6
(H) Revision of model	3	6,8
(I) Test with real data	4	8,9
(J) Final revisions	4	9,10
(K) Presentation for use	1	10,11
Dummy activities	0	3,7; 5,6

This network is re-drawn in Fig. 5.7 and the final values of the $p(i)$ are shown for each event. The minimum completion time for this project is 24 weeks and the following activities are critical: (A), (B), (E), (F), (H), (I), (J), (K).

Activity Networks 93

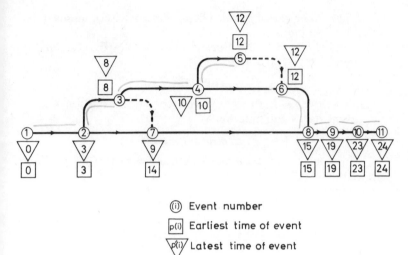

Fig. 5.7

The label $p(i)$ represents the length of the longest chain of activities from the initial event to node i and therefore the earliest time at which the event i can take place (earliest event time). There is also a latest event time by which the event *must* take place if the project is to be completed on time.

The latest event time for each node can easily be found. It is equal to the length of the critical path minus the length of the longest chain from the event in question to the terminal node. The length of this chain can be calculated using the algorithm described in this section, starting from the terminal event and labelling in the reverse direction of the arcs. Let the latest event time at node i be $p'(i)$.

Note that if event i is on the critical path $p(i) = p'(i)$, however all activities (ij) where $p(i) = p'(i)$ and $p(j) = p'(j)$ are not critical. For example in Fig. 5.7, events (4) and (6) have $p(4) = p'(4) = 10$ and $p(6) = p'(6) = 12$, but activity (4,6) is not critical.

5.6. Float

Those activities which are not on the critical path do not have to be started at their earliest start time. For example activity (7,8) in Fig. 5.7 can be started as early as week 9 (if event (7) takes place this early) and as late as week 14 without affecting the completion time of the project. The spare time permitting a job's actual start and

finish times to be varied is called float. More precisely, we make the following definitions:
(1) Total Float = Latest time of event finish − Earliest time of event start − Job duration

$$F_T = p'(j) - p(i) - a(ij)$$

(2) Free Float = Earliest time of event finish − Earliest time of event start − Job duration

$$F_F = p(j) - p(i) - a(ij)$$

For any job (ij) on the critical path $p(i) = p'(i)$, $p(j) = p'(j)$, and $p(j) - p(i) = a(ij)$. Therefore $F_F = F_T = 0$. The total float represents the maximum amount by which a given job can be displaced if all other jobs are moved to give it maximum elbow room. It might be desirable to do this with a job requiring difficult-to-obtain personnel or equipment and therefore subject to delay.

For the example which we have been treating, the float values are shown in the table below:

Activity (ij)	F_T	F_F
(1,2)	0	0
(2,3)	0	0
(2,7)	5	0
(7,8)	5	5
(3,4)	0	0
(4,5)	0	0
(4,6)	1	1
(6,8)	0	0
(8,9)	0	0
(9,10)	0	0
(10,11)	0	0

In some cases constraints may be imposed on the completion times of certain phases of the project or on the completion date of the project as a whole. When these time constraints are severe it may not be possible to complete some jobs by the specified dates. For the finish events of these jobs, we will find that $p(j) = p'(j)$, and that the total and free float take the same negative value. This negative float represents the amount by which the job must be compressed in order to meet the time constraints.

5.7. Assigning the Float

The free float represents the float that a job will have if its start and finish events take place as early as possible. Often in the initial scheduling of jobs in a project the first jobs in a given chain will be scheduled to start and finish as early as possible (thus saving their free float value). In this way the project manager obtains flexibility because if early jobs are delayed he can revise his plan assigning to them more float and cutting down on that available to later jobs. If early jobs run smoothly the float remains available in case of later trouble. As an example consider the chain formed by jobs (2,7) and (7,8). Initially job (2) would be assigned a start time of 3 and finish time of 6 (float zero). Job (7,8) could be assigned a start time of anywhere between 9 and 14 with a finish time one week later.

5.8. Resource and Manpower Levelling

As is often the case in scheduling jobs the planner is faced with conflicting objectives. We have just seen that it would be desirable to assign as much float as possible to the later jobs in order to obtain a time cushion in case of delays on the early jobs.

We have not yet however considered the availability of men and resources. When the resources required to carry out the original schedule are examined it may be found necessary to modify the schedule or in more difficult cases the logic of the network itself.

The tool which is most often used to examine the manpower and resource requirements of a given schedule is the resource profile.

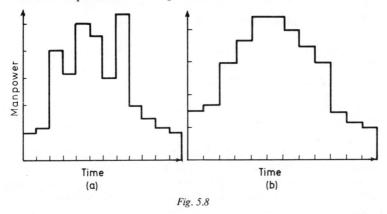

Fig. 5.8

This is a histogram showing the amount of a given resource (men of a certain trade, particular types of machinery etc.) required in each time period. An example of such a profile is given in Fig. 5.8a. In this example there are large variations in the amount of the resource required from period to period. If the resource is equipment it will be necessary to rent or purchase a quantity which will remain idle for a good part of the time. If the resource is manpower either men will be paid to remain idle or be discharged, or management will attempt to cater for the fluctuations by the hiring and firing of casual workers. If the resource is skilled or professional manpower this may not be possible.

An attempt should be made to smooth out the fluctuations. Fig. 5.8b is a resource profile that would be more satisfactory. It would provide a smooth build up and run down of the amount of resource employed as well as a fairly stable plateau in the middle time periods. Of course such a desirable profile may not be obtainable. All will depend on the float available for the various activities and on the form of the network itself. It may prove necessary to alter the network (by imposing further constraints on what jobs may be done concurrently) to smooth out the resource fluctuations even though this may increase the execution time.

The problem of resource levelling is not a simple one to deal with. It has been approached using several methods some of which are analytical (mathematical and curve fitting methods) and others which use mechanical or electrical analogues to permit a large number of possible modifications to be investigated in a reasonable time. Law and Lach[5] suggest the following heuristic rule for resource allocation:

5.9. Resource Allocation Decision Rule

Step 1 Find the subset of activities having their earliest start at the initial point ($T = 0$) in the original schedule.

Step 2 Within this subset assign the scarce resource first to the activity with the least total float and to other activities in increasing order of float until all resources have been allocated, or all jobs in the subset have received the required amount. If the entire schedule is now feasible terminate. Otherwise proceed to Step 3.

Step 3 If all resources were allocated in Step 2 the earliest start of some of the activities of the subset above will be delayed. This may in turn delay the earliest start of some subsequent activities. Adjust earliest start times and float values as necessary and proceed to Step 4.

Step 4 Find the subset of activities which have their earliest start at the time some of the scarce resource next becomes available. Return to Step 2.

5.10. Cost Curves

Associated with each job in a project are certain direct costs. These are the costs incurred for the direct labour involved and for materials used. It will often be possible to shorten job completion times by hiring more (or more competent) labour or using more expensive materials. In general lowering the completion time of a job will increase its direct cost.

Associated with the overall project are certain indirect costs or overheads such as insurance, equipment rental, building maintenance and office and supervisory staff. These costs are approximately proportional to the duration of the project and shortening overall project duration will lower indirect costs.

Overall project duration can only be lowered by lowering the completion of one or more jobs on the critical path. If the duration of the critical path is reduced sufficiently, some other path will become critical and form that point both will have to be reduced simultaneously if project duration is to be further reduced. A method of balancing direct against indirect costs can be used to find the overall project duration which minimises total cost. This can be done by examining the curves showing direct and indirect cost

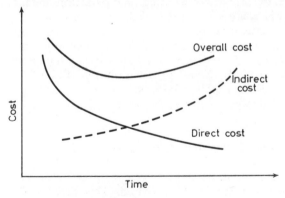

Fig. 5.9. Project cost as a function of duration

as a function of project duration. The curve showing indirect costs is not difficult to determine and a typical one is shown in Fig. 5.9.

Drawing the curve showing how direct costs vary requires examination of the jobs on the critical path (or paths). For each job, it is usual to estimate a normal completion time with its normal cost, and a crash completion time (the shortest time in which the job can be completed) and its higher crash cost. A linear relationship is then

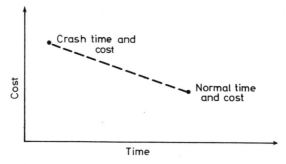

Fig. 5.10. Project cost as a function of completion time

assumed between these two points on the cost-completion time curve for that job (Fig. 5.10). The slope of the line for job (ij) is:

$$s(ij) = \frac{\text{Crash Cost} - \text{Normal Cost}}{\text{Crash Time} - \text{Normal Time}}$$

If it is desired to reduce the duration of the project by one time unit it is a simple matter to select the critical job having the smallest $s(ij)$ value. However if a reduction of two or more time units is required the problem of selecting what reductions to make on which set of jobs is a combined one. If the number of jobs considered is small, enumeration of alternatives will yield the answer. For a larger number of jobs mathematical programming may be used. In either case a curve can be drawn showing the minimum direct costs for a given overall project duration (Fig. 5.9). The curves giving direct and indirect cost can then be added to give the total cost as a function of project duration.

5.11. Uncertainty in Activity Durations—PERT

Up to this point we have treated activity durations as known quantities not subject to variation. This may be a reasonable assumption in projects of a routine nature where the activities are well known but in research and development projects, techniques which allow for the probabilistic nature of job duration must be used.

It is usual to assume that individual activity durations correspond to the Beta distribution which is defined as follows:

$$\beta(x) = \frac{x^{(p-1)}(1-x)^{(q-1)}}{B(p,q)}$$

where $B(p,q) = \int_0^1 z^{(p-1)}(1-z)^{(q-1)}dz$, and p and q are parameters which govern the skewness of the distribution.

Statistical estimates of the expected value of job duration and its standard deviation are found using optimistic, pessimistic and most likely estimates (t_o, t_p and t_m) of the duration obtained from those responsible for the activity. The estimates are made according to the following formulae:

$$\text{Job Duration estimate} = t_e = \tfrac{1}{6}(t_o + 4t_m + t_p)$$
$$\text{Standard Deviation estimate} = s_e = \tfrac{1}{6}(t_p - t_o)$$

Fig. 5.11 shows a typical Beta distribution.

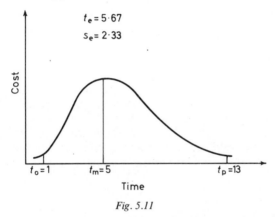

Fig. 5.11

The expected values of the job durations can be used to find the critical path through the network. However since this length is a sum of random variables it is itself a random variable. The variance of the path length is the sum of the variances of the individual job durations and by the central limit theorem, the path length is considered to be normally distributed.

As an example let us consider again the problem shown in Fig. 5.1. The critical path is of length 24 and is composed of jobs (A), (B), (E), (F), (H), (I), (J) and (K). The path containing jobs (A), (B), (E), (G), (H), (I), (J) and (K) is of length 23 and is therefore sub-critical. If we take this to be a PERT network then these lengths are known and are the means of normal distributions. Let us suppose that when calculated for the critical path the standard deviation is 3, and for the sub-critical path 5. This means that in the actual execution of the project the second path could be the critical one.

In predicting project duration it is not sufficient to consider the path with the highest expected duration. Strictly, in order to find the expected value and distribution of project completion time, it is necessary to consider all paths through the network and derive a distribution taking into account all possible interactions.

Because all paths are not considered when an estimate of project duration is made in a PERT network it can be expected that errors will occur. It has been shown that in general the estimates have an optimistic bias which may be as high as fifty or one hundred per cent. In an alternate method of estimating project duration Monté Carlo techniques are used. A specific duration is assigned to each job by sampling from its Beta distribution and the critical path found through the network. The path length is recorded (as are the job numbers) of the critical jobs and the whole process is repeated to build up a frequency histogram of the project duration. The frequency with which individual jobs are critical can also be recorded to give some idea of the probability of a given job being critical. This approach however involves high computational effort and may be too expensive in many instances.

There are methods for decomposing PERT networks into smaller networks of certain standard configurations for which probability operators have been derived[7]. Even if the network cannot completely decompose into these standard components it may be possible to reduce it in size so that Monte Carlo techniques[3,10] can be used with reasonable computational effort.

EXERCISES

1. In the table below the each activity of a project is listed in the first column, and those activities which a given job must follow are listed in the second column. From this information draw the Roy network and the CPM network for the project.

Activity	Must follow
A	—
B	M, F
C	K
D	I, L, G
E	—
F	A, C, E
G	A
H	J, B, D
I	A, C, E
J	M
K	—
L	M, F
M	K

Activity Networks

2. Carry out Exercise 1 for the project described in the following table:

Activity	Must follow
A	P, N, M
B	P
C	A, B
D	I, J, C
E	I, J, C
F	D, E
G	—
H	G
I	H
J	H, B, A
K	G
L	G
M	L
N	K, O
O	L
P	Q, O
Q	K

3. A man who is transferred from city A to city B decides to draw an activity network of his move and use it to control (as far as possible) the resulting domestic upheaval. He makes the following list:

(1) Sell house in city A.
(2) Call for bids from movers and sign a moving contract.
(3) Have belongings packed by movers.
(4) Have belongings transported to city B.
(5) Drive family to city B in car.
(6) Obtain temporary lodging in city B.
(7) Locate house in city B (assume that this can be done by an agent contacted from city A).
(8) Negotiate purchase of new house.
(9) Have belongings delivered and unpacked.
(10) Register children at new schools.
(11) Wife searches for new job (assume the search can be started by post and telephone).
(12) Call for bids on decoration of new house (this cannot start before the end of activity (11) because the wife is paying for the decoration).
(13) Have decoration done.
(14) Wind up all business in city A.
(15) Find new pub or bar and look back on the successful move over a pint of beer.

(a) Draw the activity network for this move.
(b) Number the nodes of the network using Fulkerson's algorithm.

102 *Graphs and Networks*

4. Treating the activity network of exercise 3 as a deterministic (CPM) network, the time (in days) for each activity is as follows:

Activity	Time
(1)	45
(2)	7
(3)	2
(4)	1
(5)	1
(6)	0·5
(7)	25
(8)	7
(9)	1
(10)	0·5
(11)	35
(12)	7
(13)	14
(14)	30
(15)	0·25

(a) Find the critical path through this network.
(b) Find the total float and free float for each activity.
(c) Starting from some arbitrary date draw up a work schedule for the project. (Remember this schedule would be subject to revision as the project advanced.)

5. Thinking again about his move the man of exercise 3 decides that a PERT network with probabilistic activity times would be more appropriate. His time estimates are as follows:

Activity	t_o	t_m	t_p
(1)	14	45	90
(2)	5	7	10
(3)	1	2	3
(4)	0·75	1	2
(5)	0·75	1	2
(6)	0·25	0·5	1
(7)	7	25	45
(8)	3	7	14
(9)	0·75	1	2
(10)	0·25	0·5	1
(11)	10	35	45
(12)	5	7	10
(13)	10	14	25
(14)	25	30	50
(15)	0·10	0·25	0·50

(a) Draw up a table showing the statistical estimates of job completion times and their standard deviations.
(b) Find the length of the critical path and its standard deviation.
(c) Find the longest subcritical path (second longest path through the network) and its standard deviation.

6. A prefabricated workshop building is to be erected near a road construction site for the maintenance of equipment. A portable electric generator and a 400 gallon water tank must be connected to the building. The network diagram for the project is shown

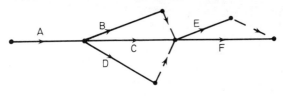

Fig. 5.12

below in Fig. 5.12. (The generator and water tank are installed at a short distance from the structure.)

The activity descriptions are listed below with the normal activity times and costs as well as the crash times and costs.

Activity	Normal		Cash	
	Time	Cost	Time	Cost
A Clear and level site	5 h	£100	3 h	£150
B Erect building	4	150	3	200
C Install generator	3	90	2	120
D Install water tank	5	25	3	75
E Connect services	2	30	2	30
F Install maintainer equipment	3	50	2	75

Assume that the entire job can be supervised by one foreman who is paid £1 per hour.
(a) Draw the curve showing how the project cost varies with project execution time.
(b) Find the critical path, with normal execution times, and at two hour intervals down to the shortest project time.
(c) Draw up a work schedule showing when activities will start and finish for each of the project execution times considered in (b).

7 (a). Assume that in the previous problem it is decided that all activities are to be carried out at normal times, that two men are

employed for each activity and that any man can be employed in any activity. Draw up a work schedule giving what you consider to be the most reasonable manpower use profile.

(b). Assume now that you are allowed to crash activity C or activity D (or both) by hiring an additional man for the job to be accelerated. Draw up a new manpower use profile and work schedule. What is the smallest work crew that can be used?

REFERENCES

1. BATTERSBY, A., *Network Analysis*, Macmillan (1964).
2. FULKERSON, D. R., 'Expected Critical Path Lengths in PERT Networks', *Operations Research*, **10**, no. 6 (1962).
3. HARTLEY, H. O. and WORTHAM, A. W., 'A Statistical Theory for PERT Critical Path Analysis', *Management Science*, **12**, no. 10 (June 1966).
4. KELLEY jr, J. E. and WALKER, M. R., *Critical Path Planning and Scheduling: An Introduction*, Mauchley Associates, Ambler Pa. (1959).
5. LAW, C. E. and LACH, D. C., *Handbook of Critical Path*, published by the Authors at Box 995, Station b, Montreal, Quebec.
6. LEVIN, R. I. and KIRKPATRICK, C. A., *Planning and Control with PERT/CPM*, McGraw-Hill (1966).
7. RINGER, L. J., 'Numerical Operators for Statistical PERT Critical Path Analysis', *Management Science*, **16**, no. 2 (B series) (1969).
8. ROY, B., 'Graphes et Ordonnancements', *Revue Francaise de Recherche Operationelle*, **25** (1962).
9. THORNLEY, G. (editor), *Critical Path Analysis in Practice*, Tavistock Publications.
10. VAN SLYKE, R. M., 'Monte-Carlo Methods and the PERT Problem', *Operations Research*, **14** (1963).

Index

Activity, network 87
  network rules 89
  precedence of 89
  uncertain duration 98
Acyclic graph 92
Acyclic network 92
AIDA 41
Algorithm, all paths 57
  critical path 92
  dynamic flow 74
  max flow 65
  min cost max flow 79
  node-numbering 90
  out-of-kilter 79
  second shortest path 53, 54
  shortest path 47, 48, 51
Application to, architecture 41
  assignment 40
  cost control 97
  construction 103
  engineering 38, 41
  games 4, 40, 41
  military transport 84
  moving problem 101
  operational research study 87, 92
  resource allocation 96, 95
  scheduling 60, 93, 94
  transportation problem 84
Arc 5, 7
  adjacent 9
  admissible 74
  dummy 23, 24, 88, 89, 92
  functions 5
  length 46
  negative length 46
  parallel 9
Assigning float 94
Assignment problem 40
Assymetric graphs and matrices 51, 52

Basic circuits 28
  co-circuits 31
  co-cycle matrix 32
  cycle matrix 30

Beta distribution 98
  function 99
Bi-partite graph 19

Capacitated nodes 70
Capacity function 62
Capacity, increase in network 80
Chain 18
  decomposition 73, 74
  definition 12
  Hamilton 2
  flow 73, 74
Circuit, definition 14
  fundamental 28
  matrix 27
  matroid 36
  Hamilton 2
  negative length 53
Co-circuit definition 14
Co-cycle compound 15
  definition 14
  dual of cycle 21
  fundamental 32
  simple 15
Co-tree, definition 16
  forming basic co-circuits with 31
  dual of tree 22
Complete graph 19
Components connected 13
Connection strong 13
Conservation equations 62
Contruction of graph from matrix 38
Cost, capacity increase 81
  curves 97
  direct 97
  indirect 97
  of project 97
Covering 40
CPM 1, 87, 91
Crash cost 97
  time 98
Critical path algorithm 92
Critical Path Method 1, 87, 91
Curve, Jordan 15

106 Index

Cut, minimum 63, 70
Cut-set 17
Cycle, compound 13
   definition 12
   detection 91
   dual of co-cycle 21
   inadmissible 90

Decanting problem 41
Decomposition of flows 73, 74
Degree, node 10
   subgraph 10
Digraph. *See* Directed graph
Dijkstra 46
   Algorithm of 46
Directed graph 5, 11
   dual of 23, 24
Distance. *See* Arc length
   matrix 51
Dominance of cycles 30
   co-cycles 32
Dreyfus 46
Dual 21
   directed 23
   matroids 36
   of a network 22
   of a two terminal network 23
   use in locating min cut 70
Duality, cycle/co-cycle 21
   tree-co-tree 22
Dynamic flow 72

Edge. *See* Arc
End points 9
Euler 1
Euler's formula 21

Faces 20
Ferrymen's problem 41
Float 93
   assigning 94
   free 94
   total 94
Flow dynamic 72
   function 62
   function modification 66
   saturating 63, 64, 69
Floyd algorithm 51
Ford algorithm 48
Ford–Fulkerson dynamic flow algorithm 74

Ford–Fulkerson min-cost/max-flow
   algorithm 79
Ford–Fulkerson out-of-kilter algorithm
   79
Ford–Fulkerson theorem 63
Forest 16
Fulkerson 90

Game 4, 40, 41
Graph, acyclic 92
   basic concepts 7
   complete 7
   connected 13
   definition 7
   directed 11
   dual 21
   faces 20
   homeomorphic 10
   imbedded 19
   isomorphic 10
   matrix representation 25
   multi-partite 10
   non-planar 19, 37
   planar 19, 37
   null 7
   one-port 17
   partial 9
   primal 21
   regions 20
   S-T planar 23
   strongly connected 13
   theory 5
   two-terminal 17
   utility 19
   disconnected 14, 17

Hall's theorem 42
Hamilton 2
Hitchcock problem 84
Hoffman algorithm 54
Homeomorphism 10

Inadmissible loop 90
Incidence matrix 25
Increasing network capacity 80
Indegree 11
Intersection of sets 34
Isomorphism 10

Jobs, assignment to 40

Index

Jordan curve 15

Kilter number 79
Knights tour 4
Königsberg bridge problem 2
Kuratowski theorem 21

Label 46, 65
Lach 96
Law 96
Link. See Arc
Longest path 92
Loop 9
   inadmissible 90

Manpower levelling 95
Matches, choosing 40
Matrices interrelation 32
   orthogonal 34
Matrix basic co-circuit 31
   basic circuit 28
   circuit 27
   co-cycle 32
   co-circuit 31
   incidence 25
   circuit 36
   co-circuit 36
   co-graphic 38
   dual 36
   graphic 38
   non-planar graph 37
   definition 34
Maximum flow algorithm 65
   min cut theorem 63
   dynamic 72, 74
   minimum cost 78
   static 65
Min-cut, location through labels 69, 70
   location through dual 70
Modulo addition 26
Monte Carlo techniques 100
Multiple sources and sinks 71
Murchland algorithm 57

Negative arc length 46
Network, activity 87
   acyclic 92
   capacity 80
   capacity increase 80
   CPM 87, 88, 91, 92

   definition 62
   directed 59, 67, 70
   distance 46
   dynamic flow 72, 73
   minimum cost flow 78
   PERT 87, 98, 99
   precedence 87
   Roy 87, 88
   static flow 62
   theory 5
   transportation 1
   two terminal 46, 62
Nodes 5, 7
   adjacent 9
   capacitated 70
   degree 10
   even 10
   numbering scheme 90
   odd 10

Ordered pair 11
Orthogonality 34
Outdegree 11

Path, critical 91
   definition 13
   flow augmenting 65
   longest 92
   second shortest 54
   second shortest 53
   shortest, all node pairs 50
   shortest 46
   all in network 57
Pavley algorithm 54
PERT 1, 87, 98
Pollack algorithm 53
Potentials, method of 87
Precedence diagrams 87
Primal graph 21
Project cost 97, 98
Project duration 99, 100

Rail network 3
Regions 20
Resource allocation 96
   levelling 95
Ring sum 26
   sum-difference 30
Road network 2
Row independence 26
Roy, method of potentials 87

Self loop 9
Sink node 17
 of flow 62
 multiple 71
Source node 17
 of flow 62
 multiple 71
Static flow 62
Strong connection 13
Subgraph 7
 even 10
 odd 10
Symmetrical graph matrix 51, 52

Tableau, all paths 58
 dynamic flow 75
 max-flow 66
 shortest path 49, 50, 51
Terminal, graph 17
 network 46, 62
Theorem, chain-cut crossings 18
 cycle/co-cycle duality 21
 Euler 4
 Ford–Fulkerson 63
 Hamilton 4
 Kuratowski's 21

max-flow/min-cut 63
 Menger's 84
 Philip Hall's 42
Time, arc traversal 74
Tournaments 41
Transportation problem 84
Travelling salesman problem 60
Tree definition 16
 dual of co-tree 22
 forming basic circuits with 28
 spanning 16
Tutte 38
Two terminal network 46, 62

Uncertainty activity durations 98
Unicursal graph 2
Union of sets 34
Utility graph 19

Value of flow 62
Vertex. *See* Node

Warehouse supply 84
Weights, lengths of arcs 46
Wollmer 71